Brooklands Books

☆ A BROOKLANDS ☆
'ROAD TEST' LIMITED EDITION

LOTUS ELITE
1957-1964

Compiled by
R.M.Clarke

ISBN 1 85520 5475

Brooklands Books
BROOKLANDS BOOKS LTD.
P.O. BOX 146, COBHAM,
SURREY, KT11 1LG. UK
sales@brooklands-books.com

A-LTETX1 — Printed in Hong Kong

Brooklands Books

MOTORING
B.B. ROAD TEST SERIES
Abarth Gold Portfolio 1950-1971
AC Ace & Aceca 1953-1983
Alfa Romeo Giulietta Gold Portfolio 1954-1965
Alfa Romeo Giulia Coupés 1963-1976
Alfa Romeo Giulia Coupés Gold Port. 1963-1976
Alfa Romeo Spider 1966-1990
Alfa Romeo Spider Gold Portfolio 1966-1991
Alfa Romeo Alfasud 1972-1984
Alfa Romeo Alfetta Gold Portfolio 1972-1987
Alfa Romeo Alfetta GTV6 1980-1986
Alvis Gold Portfolio 1919-1967
AMX & Javelin Muscle Portfolio 1968-1974
Armstrong Siddeley Gold Portfolio 1945-1960
Aston Martin Gold Portfolio 1948-1971
Aston Martin Gold Portfolio 1972-1985
Aston Martin Gold Portfolio 1985-1995
Audi Quattro Gold Portfolio 1980-1991
Audi Quattro Takes On The Competition
Austin A30 & A35 1951-1962
Austin-Healey 100 & 100/6 Gold Port. 1952-1959
Austin-Healey 3000 Ultimate Portfolio 1959-1967
Austin-Healey Sprite Gold Portfolio 1958-1971
Berkeley Sportscars Limited Edition
BMW 6 & 8 Cyl. Cars Limited Edition 1935-1960
BMW 1600 Collection No. 1 1966-1981
BMW 2002 Gold Portfolio 1968-1976
BMW 6 Cylinder Coupés & Saloons Gold P. 1969-1976
BMW 316, 318, 320 (4 cyl.) Gold Port. 1975-1990
BMW 320, 323, 325 (6 cyl.) Gold Port. 1977-1990
BMW 3 Series Gold Portfolio 1991-1997
BMW 5 Series Gold Portfolio 1981-1987
BMW 5 Series Gold Portfolio 1988-1995
BMW 6 Series Gold Portfolio 1976-1989
BMW 7 Series Performance Portfolio 1977-1986
BMW Alpina Performance Portfolio 1967-1987
BMW Alpina Performance Portfolio 1988-1998
BMW M Series Gold Portfolio 1976-1997
BMW Z3 & Z3M Limited Edition
Borgward Isabella Limited Edition
Bricklin Gold Portfolio 1974-1975
Bristol Cars 1946-2000
Buick Performance Portfolio 1947-1962
Buick Muscle Portfolio 1963-1973
Buick Riviera Performance Portfolio 1963-1978
Cadillac Allanté 1986-1993
Cadillac Automobiles 1949-1959
Cadillac Automobiles 1960-1969
Cadillac Eldorado Performance Portfolio 1967-1978
Checker Limited Edition
Chevrolet 1955-1957
Impala & SS Muscle Portfolio 1958-1972
Corvair Performance Portfolio 1959-1969
El Camino & SS Muscle Portfolio 1959-1987
Chevy II & Nova SS Muscle Portfolio 1962-1974
Chevelle & SS Muscle Portfolio 1964-1972
Caprice Limited Edition 1965-1976
Chevrolet Muscle Cars 1966-1971
Chevy Blazer 1969-1981
Camaro Muscle Portfolio 1967-1973
Chevrolet Camaro & Z-28 1973-1981
High Performance Camaros 1982-1988
Camaro Performance Portfolio 1993-2000
Chevrolet Corvette Gold Portfolio 1953-1962
Chevrolet Corvette Sting Ray Gold Port. 1963-1967
Chevrolet Corvette Gold Portfolio 1968-1977
High Performance Corvettes 1983-1989
Chrysler 300 Gold Portfolio 1955-1970
Valiant 1960-1962
Citroen Traction Avant Gold Portfolio 1934-1957
Citroen 2CV Ultimate Portfolio 1948-1990
Citroen DS & ID 1955-1975
Citroen DS & ID Gold Portfolio 1955-1975
Citroen SM 1970-1975
Shelby Cobra Gold Portfolio 1962-1969
Cobras & Cobra Replicas Gold Portfolio 1962-1989
Crosley & Crosley Specials Limited Edition
Cunningham Automobiles 1951-1955
Datsun Roadsters Performance Portfolio 1960-71
Datsun 240Z & 260Z Gold Portfolio 1970-1978
Datsun 280Z & ZX 1975-1983
DeLorean Gold Portfolio 1977-1995
De Soto Limited Edition 1952-1960
Charger Muscle Portfolio 1966-1974
Dodge Viper Performance Portfolio 1990-1998
ERA Gold Portfolio 1934-1994
Excalibur Collection No.1 1952-1981
Facel Vega 1954-1964
Ferrari Limited Edition 1947-1957
Ferrari Limited Edition 1958-1963
Ferrari Dino 308 & Mondial Gold Portfolio 1974-1985
Ferrari 328 348 Mondial Ultimate Portfolio 1986-94
Fiat 500 Gold Portfolio 1936-1972
Fiat 600 & 850 Gold Portfolio 1955-1972
Fiat Pininfarina 124 & 2000 Spider 1968-1985
Fiat X1/9 Gold Portfolio 1973-1989
Fiat Abarth Performance Portfolio 1972-1987
Ford Consul, Zephyr, Zodiac Mk. I & II 1950-1962
Ford Zephyr, Zodiac, Executive Mk. III & IV 1962-1971
Ford Cortina 1600E & GT 1967-1970
High Performance Capris Gold Portfolio 1969-1987
Capri Muscle Portfolio 1974-1987
High Performance Fiestas 1979-1991
Ford Escort RS & Mexico Limited Edition 1970-1979
High Performance Escorts Mk. I 1968-1974
High Performance Escorts Mk. II 1975-1980
High Performance Escorts 1980-1985
High Performance Escorts 1985-1990
High Perf. Sierras & Merkurs Gold Port. 1983-1990
Ford Thunderbird Performance Portfolio 1955-1957
Ford Thunderbird Performance Portfolio 1958-1963
Ford Thunderbird Performance Portfolio 1964-1976
Ford Automobiles 1949-1959
Ford Fairlane Performance Portfolio 1955-1970
Ford Ranchero Muscle Portfolio 1957-1979
Edsel Limited Edition 1957-1960
Falcon Performance Portfolio 1960-1970
Ford Galaxie & LTD Limited Edition 1960-1973
Ford GT40 Gold Portfolio 1964-1987
Mustang Muscle Portfolio 1967-1973
Ford Torino Limited Edition 1968-1974
Ford Bronco 4x4 Performance Portfolio 1966-1977
Ford Bronco 1978-1988
Goggomobil Limited Edition
Holden 1948-1962
Honda S500 • S600 • S800 Limited Edition 1962-1970
Honda CRX 1983-1987
International Scout Gold Portfolio 1961-1980
Isetta Gold Portfolio 1953-1964
ISO & Bizzarrini Gold Portfolio 1962-1974
Jaguar and SS Gold Portfolio 1931-1951
Jaguar C-Type & D-Type Gold Portfolio 1951-1960
Jaguar XK120, 140, 150 Gold Portfolio 1948-1960
Jaguar Mk. VII, VIII, IX, X, 420 Gold Port. 1950-1970
Jaguar Mk. 1 & Mk. 2 Gold Portfolio 1959-1969
Jaguar E-Type Gold Portfolio 1961-1971
Jaguar E-Type V-12 1971-1975
Jaguar S-Type & 420 Limited Edition 1963-1968
Jaguar XJ12, XJ5.3, V12 Gold Portfolio 1972-1990
Jaguar XJ6 Series I & II Gold Portfolio 1968-1979
Jaguar XJ6 Series III Perf. Portfolio 1979-1986
Jaguar XJ6 Gold Portfolio 1986-1994
Jaguar XJS Gold Portfolio 1975-1988
Jaguar XJ-S V12 Ultimate Portfolio 1988-1996
Jaguar XK8 Limited Edition
Jeep CJ-5 & CJ-6 1960-1976
Jeep CJ-5 & CJ-7 4x4 Perf. Portfolio 1976-1986
Jeep Wagoneer Performance Portfolio 1963-1991
Jeep J-Series Pickups 1970-1982
Jeepster & Commando Limited Edition 1967-1973
Jeep Cherokee & Comanche Pickups P. P. 1984-91
Jeep Wrangler 4x4 Performance Portfolio 1987-99
Jeep Cherokee & Grand Cherokee 4x4 P. P. 1992-98
Jensen Interceptor Gold Portfolio 1966-1986
Jensen - Healey Limited Edition 1972-1976
Kaiser - Frazer Limited Edition 1946-1955
Lagonda Gold Portfolio 1919-1964
Lamborghini Countach & Urraco 1974-1980
Lamborghini Countach & Jalpa 1980-1985
Lancia Aurelia & Flaminia Gold Portfolio 1950-1970
Lancia Fulvia Gold Portfolio 1963-1976
Lancia Beta Gold Portfolio 1972-1984
Lancia Delta Gold Portfolio 1979-1994
Lancia Stratos 1972-1985
Land Rover Series I 1948-1958
Land Rover Series II & IIa 1958-1971
Land Rover Series III 4x4 Perf. Portfolio 1971-1985
Land Rover 90 110 Defender Gold Portfolio 1983-1994
Land Rover Discovery 1989-1994
Land Rover Story Part One 1948-1971
Fifty Years of Selling Land Rover
Lincoln Gold Portfolio 1949-1960
Lincoln Continental Performance Portfolio 1961-1969
Lincoln Continental 1969-1976
Lotus Sports Racers Portfolio - covering 1951-1965
Lotus Seven Gold Portfolio 1957-1973
Lotus Elite Limited Edition 1957-1964
Lotus Elan Ultimate Portfolio 1962-1974
Lotus Elan & SE 1989-1992
Lotus Europa Gold Portfolio 1966-1975
Lotus Elite & Eclat 1974-1982
Lotus Elise Limited Edition
Marcos Coupés & Spyders Gold Portfolio 1960-1997
Matra Limited Edition 1965-1983
Mazda Miata MX-5 Performance Portfolio 1989-1997
Mazda Miata MX-5 Takes On The Competition
Mazda RX-7 Gold Portfolio 1978-1991
McLaren F1 Sportscar Limited Edition
Mercedes 190 & 300 SL 1954-1963
Mercedes G-Wagen 1981-1994
Mercedes S & 600 1965-1972
Mercedes S Class 1972-1979
Mercedes 230 • 250 • 280SL Gold Portfolio 1963-1971
Mercedes SLs & SLCs Gold Portfolio 1971-1989
Mercedes SLs Performance Portfolio 1989-1994
Mercury Limited Edition 1947-1959
Mercury Comet & Cyclone Limited Edition 1960-1970
Cougar Limited Edition 1967-1973
Messerschmitt Gold Portfolio 1954-1964
MG Gold Portfolio 1929-1939
MG TA & TC Gold Portfolio 1936-1949
MG TD & TF Gold Portfolio 1949-1955
MGA & Twin Cam Gold Portfolio 1955-1962
MG Midget Gold Portfolio 1961-1979
MGB Roadsters 1962-1980
MGB MGC & V8 Gold Portfolio 1962-1980
MGB GT 1965-1980
MGC & MGB GT V8 Limited Edition
MG Y-Type & Magnette ZA/ZB Limited Edition
MGF Limited Edition
Mini Gold Portfolio 1959-1969
Mini Gold Portfolio 1969-1980
Mini Gold Portfolio 1981-1997
High Performance Minis Gold Portfolio 1960-1973
Mini Cooper Gold Portfolio 1961-1971
Mini Moke Gold Portfolio 1964-1994
Morgan Three-Wheeler Gold Portfolio 1910-1952
Morgan Plus 4 & Four 4 Gold Portfolio 1936-1967
Morris Minor Collection No. 1 1948-1980
Shelby Mustang Muscle Portfolio 1965-1970
High Performance Mustang IIs 1974-1978
Mustang 5.0L Muscle Portfolio 1982-1993
Mustang 5.0L Takes On The Competition
Nash & Nash-Healey Limited Edition 1949-1957
Nash-Austin Metropolitan Gold Portfolio 1954-1962
NSU Ro80 Limited Edition
NSX Performance Portfolio 1989-1999
Oldsmobile Automobiles 1955-1963
Oldsmobile Muscle Portfolio 1964-1971
Cutlass & 4-4-2 Muscle Portfolio 1964-1974
Oldsmobile Toronado 1966-1978
Opel GT Gold Portfolio 1968-1973
Opel Manta Limited Edition 1970-1975
Packard Gold Portfolio 1946-1958
Pantera Gold Portfolio 1970-1989
Panther Gold Portfolio 1972-1990
Barracuda Muscle Portfolio 1964-1974
Pontiac Limited Edition 1949-1960
Pontiac Tempest & GTO 1961-1965
GTO Muscle Portfolio 1964-1974
Firebird & Trans-Am Muscle Portfolio 1967-1972
Firebird & Trans-Am Muscle Portfolio 1973-1981
High Performance Firebirds 1982-1988
Firebird & Trans Am Performance Portfolio 1993-2000
Pontiac Fiero Performance Portfolio 1984-1988
Porsche 356 Gold Portfolio 1953-1965
Porsche 912 Limited Edition
Porsche 911 1965-1969
Porsche 911 1970-1972
Porsche 911 1973-1977
Porsche 911 SC & Turbo Gold Portfolio 1978-1983
Porsche 911 Carrera & Turbo Gold Port. 1984-1989
Porsche 911 Gold Portfolio 1990-1997
Porsche 911 Takes On The Competition 1990-1997
Porsche 914 Ultimate Portfolio
Porsche 924 Gold Portfolio 1975-1988
Porsche 928 Performance Portfolio 1977-1994
Porsche 928 Takes On The Competition
Porsche 944 Gold Portfolio 1981-1991
Porsche 968 Limited Edition
Porsche Boxster Limited Edition
Railton & Brough Superior Gold Portfolio 1933-1950
Range Rover Gold Portfolio 1970-1985
Range Rover Gold Portfolio 1986-1995
Range Rover Takes on the Competition
Reliant Scimitar 1964-1986
Renault Alpine Gold Portfolio 1958-1994
Riley Gold Portfolio 1924-1939
R. R. Silver Cloud & Bentley 'S' Series Gold P. 1955-65
Rolls Royce Silver Shadow Ultimate Portfolio 1965-80
Rolls Royce & Bentley Gold Portfolio 1980-1989
Rolls Royce & Bentley Limited Edition 1990-1997
Rover P4 1949-1959
Rover 3 & 3.5 Litre Gold Portfolio 1958-1973
Rover 2000 & 2200 1963-1977
Rover 3500 & Vitesse 1976-1986
Saab Sonett Collection No.1 1966-1974
Saab Turbo 1976-1983
Studebaker Gold Portfolio 1947-1966
Studebaker Hawks & Larks 1956-1963
Avanti 1962-1990
Suzuki SJ Gold Portfolio 1971-1997
Vitara, Sidekick & Geo Tracker Perf. Port. 1988-1997
Sunbeam Tiger & Alpine Gold Portfolio 1959-1967
Toyota Land Cruiser Gold Portfolio 1956-1987
Toyota Land Cruiser 1988-1997
Toyota MR2 Gold Portfolio 1984-1997
Toyota MR2 Takes On The Competition
Triumph TR2 & TR3 Gold Portfolio 1952-1961
Triumph TR4, TR5, TR250 1961-1968
Triumph TR6 Gold Portfolio 1969-1976
Triumph TR7 & TR8 Gold Portfolio 1975-1982
Triumph Herald 1959-1971
Triumph Vitesse 1962-1971
Triumph Spitfire Gold Portfolio 1962-1980
Triumph 2000, 2.5, 2500 1963-1977
Triumph GT6 Gold Portfolio 1966-1974
Triumph Stag Gold Portfolio 1970-1977
Triumph Dolomite Sprint Limited Edition
TVR Gold Portfolio 1959-1986
TVR Performance Portfolio 1986-1994
VW Beetle Gold Portfolio 1935-1967
VW Beetle Gold Portfolio 1968-1991
VW Beetle Collection No.1 1970-1982
VW Karmann Ghia 1955-1982
VW Bus, Camper, Van 1954-1967
VW Bus, Camper, Van Perf. Portfolio 1968-1979
VW Bus, Camper, Van 1979-1989
VW Scirocco 1974-1981
Volvo PV444 & PV544 1945-1965
Volvo 120 Amazon Ultimate Portfolio
Volvo 1800 Gold Portfolio 1960-1973
Volvo 140 & 160 Series Gold Portfolio 1966-1975
Forty Years of Selling Volvo
Westfield Limited Edition

B.B. ROAD & TRACK SERIES
Road & Track on Alfa Romeo 1964-1970
Road & Track on Alfa Romeo 1971-1976
Road & Track on Aston Martin 1962-1990
R & T on Auburn Cord and Duesenburg 1952-84
Road & Track on Audi & Auto Union 1952-1980
Road & Track on Audi & Auto Union 1980-1986
Road & Track on Austin Healey 1953-1970
Road & Track on BMW Cars 1966-1974
Road & Track on BMW Cars 1975-1978
Road & Track on BMW Cars 1979-1983
R & T on Cobra, Shelby & Ford GT40 1962-1992
Road & Track on Corvette 1953-1967
Road & Track on Corvette 1968-1982
Road & Track on Corvette 1982-1986
Road & Track on Corvette 1986-1990
Road & Track on Ferrari 1975-1981
Road & Track on Ferrari 1981-1984
Road & Track on Ferrari 1984-1988
Road & Track on Fiat Sports Cars 1968-1987
Road & Track on Jaguar 1950-1960
Road & Track on Jaguar 1961-1968
Road & Track on Jaguar 1968-1974
Road & Track on Jaguar 1974-1982
Road & Track on Jaguar 1983-1989
Road & Track on Lamborghini 1964-1985
Road & Track on Lotus 1972-1983
R & T on Mazda RX-7 & MX-5 Miata 1986-1991
Road & Track on Mercedes 1952-1962
Road & Track on Mercedes 1963-1970
Road & Track on Mercedes 1971-1979
Road & Track on Mercedes 1980-1987
Road & Track on MG Sports Cars 1949-1961
Road & Track on MG Sports Cars 1962-1980
R & T on Nissan 300-ZX & Turbo 1984-1989
Road & Track on Pontiac 1960-1983
Road & Track on Porsche 1951-1967
Road & Track on Porsche 1968-1971
Road & Track on Porsche 1972-1975
Road & Track on Porsche 1975-1978
Road & Track on Porsche 1979-1982
Road & Track on Porsche 1985-1988
R & T on Rolls Royce & Bentley 1950-1965
R & T on Rolls Royce & Bentley 1966-1984
Road & Track on Saab 1972-1992
R & T on Toyota Sports & GT Cars 1966-1984
R & T on Triumph Sports Cars 1953-1967
R & T on Triumph Sports Cars 1967-1974
R & T on Triumph Sports Cars 1974-1982
Road & Track on Volkswagen 1951-1968
Road & Track on Volkswagen 1968-1978
Road & Track on Volkswagen 1978-1985
Road & Track on Volvo 1957-1974
Road & Track on Volvo 1977-1994
Road & Track - Henry Manney at Large & Abroad
Road & Track - Peter Egan "At Large"
Road & Track - Best of PS

B.B. CAR AND DRIVER SERIES
Car and Driver on BMW 1955-1977
Car and Driver on Corvette 1978-1982
Car and Driver on Corvette 1983-1988
C and D on Datsun Z 1600 & 2000 1966-1984
Car and Driver on Ferrari 1955-1962
Car and Driver on Ferrari 1963-1975
Car and Driver on Ferrari 1976-1983
Car and Driver on Mopar 1956-1967
Car and Driver on Mustang 1964-1972
Car and Driver on Pontiac 1961-1975
Car and Driver on Porsche 1955-1962
Car and Driver on Porsche 1963-1970
Car and Driver on Porsche 1970-1976
Car and Driver on Porsche 1977-1981
Car and Driver on Porsche 1982-1986
Car and Driver on Volvo 1955-1986

RACING & THE LAND SPEED RECORD
The Land Speed Record 1898-1919
The Land Speed Record 1920-1929
The Land Speed Record 1930-1939
The Land Speed Record 1940-1962
The Land Speed Record 1963-1999
The Land Speed Record 1898-1999
Can-Am Racing 1966-1969
Can-Am Racing 1970-1974
Can-Am Racing Cars 1966-1974
Can-Am Racing Portfolio 1966-1974
The Carrera Panamericana Mexico - 1950-1954
Le Mans - The Bentley & Alfa Years - 1923-1939
Le Mans - The Jaguar Years - 1949-1957
Le Mans - The Ferrari Years - 1958-1965
Le Mans - The Ford & Matra Years - 1966-1974
Le Mans - The Porsche Years - 1975-1982
Le Mans - The Porsche & Jaguar Years - 1983-91
Le Mans - The Porsche & Peugeot Years - 1992-99
Mille Miglia - The Alfa & Ferrari Years - 1927-1951
Mille Miglia - The Ferrari & Mercedes Years - 1952-57
Targa Florio - The Post War Years - 1948-1973
Targa Florio - The Porsche & Ferrari Years - 1955-1964
Targa Florio - The Porsche Years - 1965-1973

B.B. PRACTICAL CLASSICS SERIES
PC on Austin A40 Restoration
PC on Land Rover Restoration
PC on Midget/Sprite Restoration
PC on MGB Restoration
PC on Sunbeam Rapier Restoration
PC on Triumph Herald/Vitesse

B.B. HOT ROD 'ENGINE' SERIES
Chevy 265 & 283
Chevy 302 & 327
Chevy 348 & 409
Chevy 350 & 400
Chevy 396 & 427
Chevy 454 thru 512
Chrysler Hemi
Chrysler 273, 318, 340 & 360
Chrysler 361, 383, 400, 413, 426 & 440
Ford 289, 302, Boss 302 & 351W
Ford 351C & Boss 351
Ford Big Block

B.B. RESTORATION & GUIDE SERIES
BMW 2002 - A Comprehensive Guide
BMW '02 Restoration Guide
Classic Camaro Restoration
Chevrolet High Performance Tips & Techniques
Chevy Engine Swapping Tips & Techniques
Chevy-GMC Pickup Repair
Engine Swapping Tips & Techniques
Land Rover Restoration Portfolio
MG 'T' Series Restoration Guide
MGA Restoration Guide
Mustang Restoration Tips & Techniques
The Great Classic Muscle Cars Compared

MOTORCYCLING
B.B. ROAD TEST SERIES
AJS & Matchless Gold Portfolio 1945-1966
BMW Motorcycles Gold Portfolio 1950-1971
BMW Motorcycles Gold Portfolio 1971-1976
BSA Singles Gold Portfolio 1945-1963
BSA Singles Gold Portfolio 1964-1974
BSA Twins A7 & A10 Gold Portfolio 1946-1962
BSA Twins A50 & A65 Gold Portfolio 1962-1973
BSA & Triumph Triples Gold Portfolio 1968-1976
Ducati Gold Portfolio 1960-1973
Ducati Gold Portfolio 1974-1978
Ducati Gold Portfolio 1978-1982
Harley-Davidson Sportsters Pref. Port. 1965-1976
Harley-Davidson Super Glide Perf. Port. 1971-1981
Harley-Davidson FXR Series Perf. Port. 1982-1992
Honda CB750 Gold Portfolio 1969-1978
Honda CB500 & 550 Fours Perf. Port. 1971-1977
Honda CB350 & 400 Fours Perf. Port. 1972-1978
Honda Gold Wing Gold Portfolio 1975-1995
Honda CBX 1000 Gold Portfolio 1978-1982
Honda RC30 Performance Portfolio 1988-1992
Kawasaki Z1 900 Performance Portfolio 1972-1977
Kawasaki 500 & 750 Triples Perf. Port. 1969-1976
Kawasaki GPZ 900R Performce Port. 1984-1996
Laverda Gold Portfolio 1967-1977
Laverda Performance Portfolio 1978-1988
Laverda Jota Performance Portfolio 1976-1985
Moto Guzzi Gold Portfolio 1949-1973
Moto Guzzi Le Mans Performance Portfolio 1976-89
Norton Commando Gold Portfolio 1968-1977
Suzuki GT750 Performance Portfolio 1971-1977
Suzuki GS1000 Performance Portfolio 1978-1981
Triumph Bonneville Gold Portfolio 1959-1983
Vincent Gold Portfolio 1945-1980
Yamaha RD350/400 Performance Portfolio 1972-79

B.B. CYCLE WORLD SERIES
Cycle World on BMW 1974-1980
Cycle World on BMW 1981-1986
Cycle World on Ducati 1982-1991
Cycle World on Harley-Davidson 1978-1983
Cycle World on Harley-Davidson 1983-1987
Cycle World on Harley-Davidson 1987-1990
Cycle World on Harley-Davidson 1990-1992
Cycle World on Honda 1962-1967
Cycle World on Honda 1968-1971
Cycle World on Honda 1971-1974
Cycle World on Husqvarna 1966-1976
Cycle World on Husqvarna 1977-1984
Cycle World on Kawasaki 1966-1971
Cycle World on Kawasaki Off-Road Bikes 1972-1979
Cycle World on Kawasaki Street Bikes 1972-1976
Cycle World on Norton 1962-1971
Cycle World on Suzuki 1962-1970
Cycle World on Suzuki Off-Road Bikes 1971-1976
Cycle World on Suzuki Street Bikes 1971-1976
Cycle World on Triumph 1967-1972
Cycle World on Yamaha 1962-1969
Cycle World on Yamaha Off-Road Bikes 1970-1974
Cycle World on Yamaha Street Bikes 1970-1974

MILITARY
VEHICLES & AEROPLANES
Complete WW2 Military Jeep Manual
Dodge WW2 Military Portfolio 1940-1945
German Military Equipment WW2
Hail To The Jeep
Military & Civilian Amphibians 1940-1990
Off Road Jeeps Civilian & Military 1944-1971
Silhouette Handbook of US Army Air Forces Aeroplanes
US Military Vehicles 1941-1945
US Army Military Vehicles WW2-TM9-2800
VW Kubelwagen Military Portfolio 1940-1990
WW2 Allied Vehicles Military Portfolio 1939-1945
WW2 Jeep Military Portfolio 1941-1945

10/8Z0

CONTENTS

5	For the Elite	*Sports Car Illustrated (UK)*	Nov		1957
7	Star Car - Lotus Elite	*Autosport*	Oct	25	1957
8	Lotus Elite New Cars Described	*Autocar*	Oct	18	1957
10	A New Style of Lotus	*Sports Cars Illustrated (US)*	Jan		1958
12	The Elite Road Impressions	*Sports Car & Lotus Owner*	Nov		1958
14	On the Road with the Lotus Elite	*Sports Cars Illustrated (UK)*	Dec		1958
16	First Test Run of the Lotus	*Motor Trend*	Feb		1959
19	Lotus Elite Road Test	*Road & Track*	Jan		1960
22	Lotus Elite Road Test	*Sports Cars Illustrated (US)*	June		1960
24	The Lotus Elite Road Test	*Motor*	May	11	1960
30	Lotus Elite Technical Appraisal	*Motor*	June	22	1960
34	The Lotus Elite Road Impressions	*Motor Racing*	July		1960
36	Lotus Elite A First Long Look	*Sports Car World*	July		1960
39	Lotus Elite	*Sports Car Graphic*	May		1960
40	Lotus Elite Road Test	*Sports Car Graphic*	Aug		1961
45	The Lotus Elite - LOV 1 Road Test	*Autosport*	Feb	24	1961
46	The 'Do-It-Yourself' Lotus Elite Road Test	*Austosport*	Dec	1	1961
48	An Elite in the Family	*Sporting Motorist*	Jan		1962
54	Three Months with the Elite Case History	*Sporting Motorist*	Mar		1962
56	The Lotus That Came in a Crate	*Car and Driver*	July		1962
60	The Lotus Elite Re-assessment	*Autosport*	Mar	2	1962
62	Lotus Elite (Special Equipment)	*Motor Annual*			1963
65	Lotus Elite S/E Road Test	*Small Car*	Dec		1962
69	Lotus Elite Road & Track Test	*Motor Sport*	Dec		1962
72	Lotus Elite Road Test	*Road & Track*	Apr		1963
76	Lotus Elite Series II	*World Car Catalogue*			1964
77	Lotus Elite Road Test	*Cars Illustrated*	Aug		1964
79	Lotus Elite An Appreciation	*Road & Track*	June		1964
80	Charge of the Light Brigade	*Classic Cars*	Mar		1998
86	Featherweight Champion Profile	*Classic & Sports Car*	Apr		1988

Brooklands Books

ACKNOWLEDGEMENTS

For more than 35 years, Brooklands Books have been publishing compilations of road tests and other articles from the English speaking world's leading motoring magazines. We have already published more than 600 titles, and in these we have made available to motoring enthusiasts some 20,000 stories which would otherwise have become hard to find. For the most part, our books focus on a single model, and as such they have become an invaluable source of information. As Bill Boddy of *Motor Sport* was kind enough to write when reviewing one of our Gold Portfolio volumes, the Brooklands catalogue "must now constitute the most complete historical source of reference available, at least of the more recent makes and models."

Even so, we are constantly being asked to publish new titles on cars which have a narrower appeal than those we have already covered in our main series. The economics of book production make it impossible to cover these subjects in our main series, but Limited Edition volumes like this one give us a way to tackle these less popular but no less worthy subjects. This additional range of books is matched by a Limited Edition - Extra series, which contains volumes with further material to supplement existing titles in our Road Test and Gold Portfolio ranges.

Both the Limited Edition and Limited Edition - Extra series maintain the same high standards of presentation and reproduction set by our established ranges. However, each volume is printed in smaller quantities - which is perhaps the best reason we can think of why you should buy this book now. We would also like to remind readers that we are always open to suggestions for new titles; perhaps your club or interest group would like us to consider a book on your particular subject?

Finally, we are more than pleased to acknowledge that Brooklands Books rely on the help and co-operation of those who publish the magazines where the articles in our books originally appeared. For this present volume, we gratefully acknowledge the continued support of the publishers of *Autosport, Car and Driver, Cars Illustrated, Classic & Sports Car, Classic Cars, Motor, Motor Racing, Motor Sport, Motor Trend, Road & Track, Small Car, Sporting Motorist, Sports Car & Lotus Owner, Sports Car Graphic, Sports Cars Illustrated (UK), Sports Cars Illustrated (US), Sports Car World* and *World Car Catalogue* for allowing us to include their valuable and informative copyright stories.

R.M. Clarke.

Two new models from Lotus intended
FOR THE ELITE

Aesthetic and aerodynamic perfection have been achieved in the Lotus Elite

The rear of the coupe is no less handsome than the front

RARELY has the public been shown a prototype of a new model as perfect in appearance and attractive in specification as the new Lotus Elite. Although this is an entirely original design, and moreover a departure from anything else previously attempted by its designer, Colin Chapman, the new Elite reveals a sureness of line, and harmony of proportion which, if the car were not a Lotus, one would imagine was the result of a 10-year programme of refinement on a basically good looking model.

Not only is the Elite exceptional in appearance, but in specification too. Unlike its stable-mates, the Eleven and the new Seven, which have also been introduced in the past few days, the Elite has an integral structure, in which resin-bonded glass fibre mouldings are used throughout. Unlike earlier glass fibre designs, this one utilizes a bare minimum of steel reinforcement. Front and rear suspension are derived from the Lotus F2 single seater, with independent front suspension using wishbones and coil springs and independent rear suspension using Chapman struts. The engine is a Coventry Climax single overhead camshaft four cylinder unit of 1,216 c.c. capacity, driving a four-speed gearbox identical to that of the M.G.A. through a conventional 8-in. dry clutch. Disc brakes are fitted outboard on the front wheels and inboard at the rear as on the Lotus Eleven Le Mans and F2 cars, while centre lock wire wheels are used carrying 4.90 × 15 in. tyres.

From the outset, the designer has kept in mind the need to provide every comfort possible in a two-seater of exceptionally high performance. Thus although the overall height of the car has been kept to only 3 ft. 10 in. there is ample room internally for two large people, while entry and exit are facilitated by sweeping the forward edge of the door well forward to allow even the longest legs to be inserted without acrobatics. Further by using curved glasses in the door windows, it has been possible to extend the door aperture into the roof, thus further increasing ease of entry.

Accommodation for one or two spare wheels is arranged inside the car behind the two seats, which, incidentally were specially developed for the Elite by the Dunlop organization. An external locker for luggage is reached through the lift-up door lid. A full set of instruments is mounted as standard above a full-width parcel shelf and provision is made for the installation of a radio, and either a heating and ventilating system, or an air conditioning unit if required. Ventilation has been carefully studied, and ventilation slots are provided to allow air into the car above the doors with outlets in an extension of the roof line over the rear window.

One of the advantages of the type of glass reinforced resin moulding used in the Elite is its self-damping properties. Thus drumming is impossible and the reluctance of the material to pass on road excited or engine noise should result in it being exceptionally silent. To achieve the high degree of stiffness which has been part of the secret of the phenomenal road holding qualities of earlier Lotus models, recourse has been made to a form of sandwich construction in highly stressed areas such as the roof, where the outer glass reinforced panel is bonded to a thin sheet of cellular glass. The only steel found in the structure consists of a tube running round the scuttle and around the top of the screen, bonded into the structure and serving as a tie between the roof and the door sills and scuttle, and perforated steel plates are bonded into the structure at the engine, suspension and steering gear mounting points. The attachment points themselves are welded to these plates which are then incorporated in the moulds.

Another original detail is the use of integral fuel tanks, one in each front wing providing a total capacity of 18 gallons. These tanks are formed by bulkheads arranged transversely between the outer body panels and the inner wall or wheel arch extensions, so that in addition to storing fuel they also serve as box sections stiffening the structure. Similarly at the rear, the inner wheel arch and the external valances serve as box section stiffening units, while other similar stress bearing members are the propeller shaft tunnel, the air duct to the

Stark in the vintage tradition, but entirely modern in mechanical specification, the Lotus Seven recalls the Lotus Mk. VI

Space for two spare wheels is arranged behind the seats in the Elite

By extending the doors well forward and into the roof by using curved window glasses, ease of entry is assured

radiator, and of course the floor, which is in the form of a complete undertray with openings for sump and final drive units. The doors are double skinned, and have integral arm rests. At front and rear the bumpers are blended into the line of the car, but are not moulded into the surrounding panels.

For this model, a new Coventry Climax engine has been developed. This closely resembles the well-known 1,100 c.c. engine and has the same stroke, 66.6 mm., while the bore, at 76.2 mm. is the same as that of the twin overhead camshaft 1,500 c.c. engine. Thus a capacity of 1,261 c.c. is obtained, and with its single overhead camshaft, 10:1 compression ratio, and twin S.U. H4 carburetters, it produces 83 b.h.p. at 6,300 r.p.m.

Production of the Elite will be initiated late in the spring of 1958 after the prototype has undergone a strenuous period of development testing. The price quoted for the model is £1,300 to which purchase tax of £651 must be added.

Meanwhile another new Lotus is scheduled for early production. This is the Seven, a modernised, and improved version of the Mk. VI, the model which first brought the name of Lotus into prominence. Using a multi-tubular space frame derived from the Lotus Eleven Le Mans, the Seven dispenses with the costly, but efficient aerodynamic bodywork of the Eleven and is clothed instead in traditionally stark open fashion, with cycle-type front wings, external head-lamps and exhaust system. With wishbone front suspension of the latest Lotus type, and a rigid rear axle using the familiar Lotus double trailing arms and transverse rod, the road holding and ride should be exceptionally good.

The simplicity of the body, which is equipped as standard with a full width glass windscreen, and of the structure generally should also appeal largely to the enthusiast wishing to build his own car.

All weather equipment is to be available for the Seven, and a luggage compartment enables it to be used for touring as well as for club and other competitions. Thus the current Lotus range includes cars suitable for high-speed touring and the faster Continental rallies, a Formula 2 car, and various versions of the sports-racing models. With the addition to the range of this low priced but high-performance car with a Ford 100E 1,172 c.c. engine and a three-speed gearbox, every enthusiastic motorist, from the beginner to the driver of International standard, is catered for. It is also interesting to consider that within the last year Lotus Engineering Co., has introduced an entirely new, five-speed gearbox combined with the final drive, an original and successful independent rear suspension system, and now, with the Elite, has proved itself capable of designing and producing, without recourse to Continental designers or bodybuilders, one of the best looking (if not the best looking) cars ever to come out of this country.

Colin Chapman provides a scale for the Seven and the Elite—and he is not over 5 ft. 10 in. tall

The luggage locker is roomy with the battery recessed in the left wing and the tool box in the right-hand one

6

Star Car ★ ★ ★

One of the most dramatic last-minute exhibits at the Motor Show—the 1,220 c.c. Climax-powered Lotus Elite G.T. Coupe

CONTROL LAYOUT (left) is functional, yet pleasing to the eye. The interior is fully furnished and there is a useful boot (above) with the lightweight battery stowed in its side.

PROUD MOMENT for Colin Chapman as he stands beside the waist-high coupé. The chassis-less body shell is of glass-reinforced plastic, and the suspension is all independent, with disc brakes all round.

LOTUS ELITE

NEW CARS DESCRIBED

GRAN TURISMO COUPÉ WITH 1,216 c.c. COVENTRY-CLIMAX ENGINE

IT was to be expected that when Colin Chapman decided to graduate from the rather special sports-racing car field and produce a true passenger car, he would evolve a very original design. In the same way that he has achieved a high power-weight ratio with good aerodynamic form on his very successful sports-racing cars, he has aimed at the same targets for his latest creation, a Gran Turismo coupé to be known as the Elite. As with previous designs it is a team effort—Chapman has been responsible for the structural side, the styling is the work of Peter Kirwin-Taylor, John Frayling undertook the sculptoral work on the full-scale model, and the aerodynamics have been looked after by Frank Costin, who has done much work with Colin Chapman.

Low overall weight has been achieved by the use of resin-bonded glass for the chassis-body construction, which is built up from individual mouldings. The main sections consist of a large front unit, a combined roof and rear moulding and a floor structure. There are already in existence several examples of resin-bonded glass construction, but the Elite is unusual in that the use of metal reinforcing members is eliminated almost completely.

Each of the main sections comprises several individual mouldings, the basic material of which is a polyester resin. They are joined together by an epoxide resin adhesive which provides a pure bond between the joints, and eliminates the need for auxiliary riveting. Mould thickness varies between 0.10in for general construction work and 0.20in in the stress panels. Depending on the loads to be carried, the bond is either of glass fibre or woven glass.

In some sections, such as the roof members, a foam sheet also is used; this is made of cellular glass similar to foam rubber, to each side of which is bonded a thin sheet of woven glass. This type of construction is similar to the metal honeycomb sandwich technique widely used in the aircraft industry.

The main load-carrying members at the rear form a triangular box section, merging into the rear wing valances, which provide the main attachment points for the chassis-mounted final drive unit. From these points the stressed members are carried forward in three ways. There is a deep tunnel section for the propeller shaft, a box section sill on each side, and a cellular box structure forming the cant rails in the roof and merging into another box section above the windscreen. These three members all converge in the scuttle, where the only basic metal reinforcement member is used—a steel tube bonded into the glass fibre and tying together the lower sills, scuttle structure, and stress-carrying members in the roof. On either side, the lower end of this tube forms a side jacking point.

From the scuttle there is a rectangular wing valance section which forms moulded-in fuel tanks, and these terminate in a thick cross member at the front, on which are the mounting points for the front suspension. This becomes a deep cross member which serves also as a deflector plate for the ducted radiator. The floor member—an important component in the load-carrying structure—is unbroken except for small openings beneath the engine, and the final drive. Where these openings occur thickened, turned-up flanges are provided to restore structural stiffness.

At the various attachment points for the suspension, engine mountings and steering, a special technique has been evolved. The actual attachment points are welded to a large plate of perforated steel which is bonded into the moulds, thereby spreading the local loads over a wide area into the main stress-carrying structure.

The power unit is the new Coventry Climax 1,216 c.c. FWE engine, described on page 598. In its single carburettor form it develops 72 b.h.p. at 6,100 r.p.m. Mounted directly to it is a B.M.C. B-type gear box, identical with that fitted to the M.G.A. Remote control is not needed, as the short gear shift lever is mounted directly above the selector forks. At present the engine is not fitted with a cooling fan, a ducted radiator with outlet beneath the floor being provided. If development work proves that a fan is necessary it will probably be electrically driven and thermostatically controlled.

Front and rear suspensions are the same as those fitted to the formula 2 Lotus, but wheel deflections have been increased from 6in to 7in with adjustment to spring rates and damper settings. At the front there is a tubular lower wishbone, with a single arm at the top combined with an anti-roll bar. Each wheel pivots at the top on a ball, and at the bottom on a threaded trunnion which accommodates the static weight of the car.

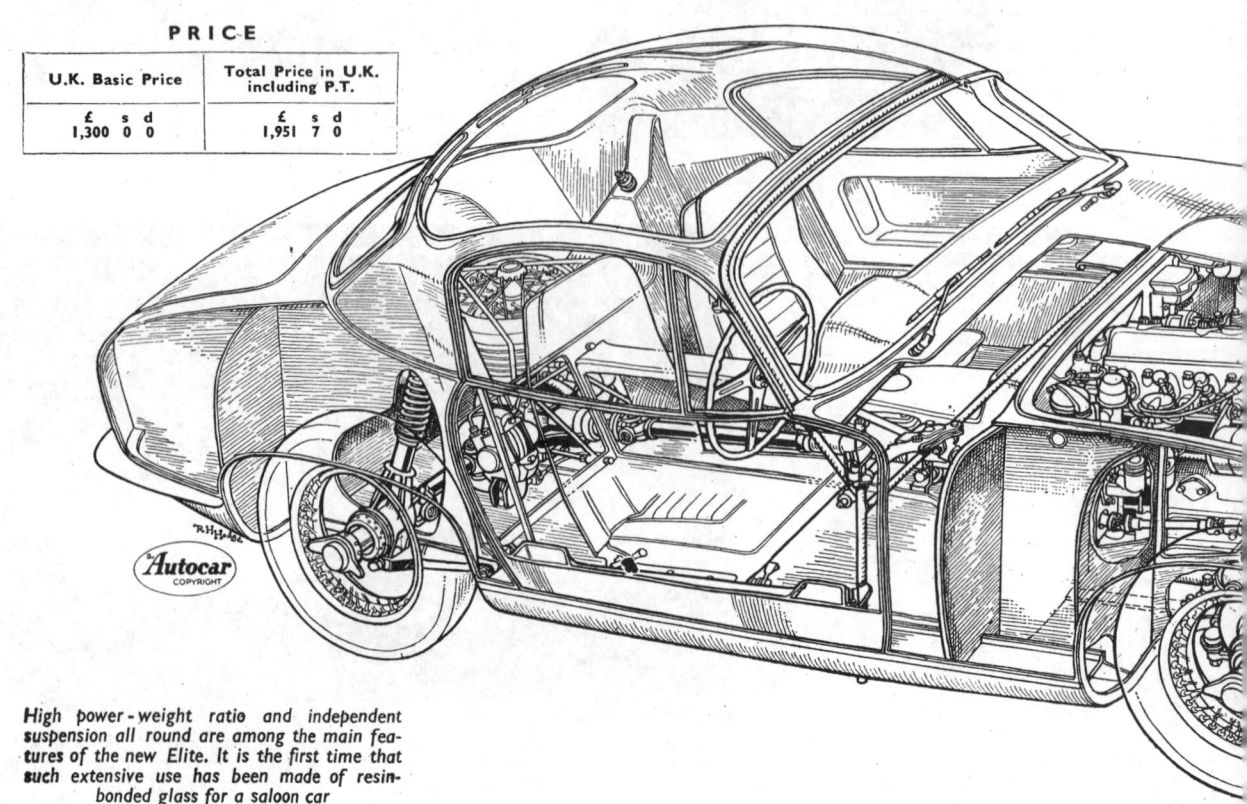

PRICE

U.K. Basic Price	Total Price in U.K. including P.T.
£ s d 1,300 0 0	£ s d 1,951 7 0

High power-weight ratio and independent suspension all round are among the main features of the new Elite. It is the first time that such extensive use has been made of resin-bonded glass for a saloon car

The combined telescopic damper and coil spring is mounted on the outboard end of the lower wishbone, with an inward inclination at the top where it is secured on an extension of the pivot bearing. The rack and pinion steering is located behind the radiator, and the connecting shaft to the steering wheel incorporates universal joints.

Rear suspension is fully independent by the Chapman strut system. It is designed to have a controlled degree of camber change with increasing load, to provide good handling characteristics under all conditions. There are three basic members—the double articulated wheel-driving shafts to locate the wheels laterally, a forward-facing arm for longitudinal location, and the strut member comprising a telescopic damper and surrounding coil spring; the lower end of the damper has an interference fit in an aluminium casting which contains the wheel hub bearings. Centre lock wheels (4.90-15in) and special high-performance tyres with nylon casings are used.

Girling 9½in diameter disc brakes are fitted, those at the rear being mounted inboard on the final drive unit; at the front they are located outboard. Each has quickly detachable pads, and the hand brake, operated by a horizontal lever beneath the dash, is connected by cable and conduits to separate mechanical calipers at the rear.

The coupé is intended only as a two-seater, and the space behind the seats is used for carrying the horizontally-mounted spare wheel, above which is a platform to provide additional luggage space; two spare wheels can be specified if required. There is at the rear a normal luggage compartment of reasonable capacity. In a separate compartment on the left-hand side of the luggage boot is the battery, and matching it on the right a compartment for tools.

In the instrument panel are mounted a speedometer and a rev counter, with flanking dials for ammeter, engine temperature and oil pressure. To the left of the main instruments is a row of five identical switches, with the light switch and trafficator on the right-hand side of the panel. Beneath the facia is a full-width parcels shelf, with the heater above.

In the box section cant rails and roof section are ventilation slots communicating with the inside of the car and with exit slots above the rear window, so that a controlled, draught-free air flow is obtained. The framed door windows are fixed, with a hinged front quarter-light in each. The doors have wide openings and are of double-skin construction to provide rigidity, a recessed moulding forming an arm rest on each side. The front bumper is integral with the air intake and there is a control for adjusting the head lamps (in which are incorporated the side lights) from within the car.

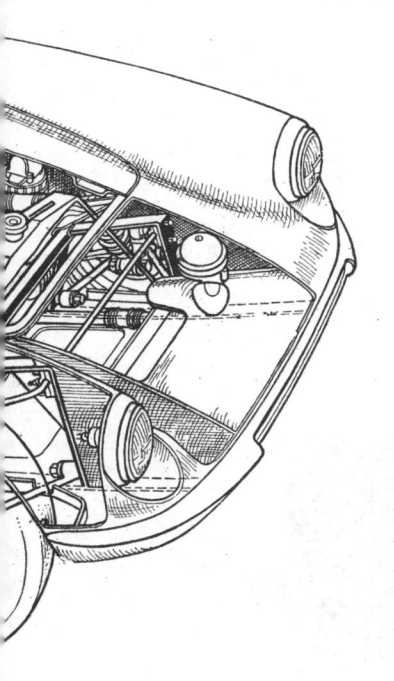

Above: Clean, crisp lines with good all-round visibility and calculated aerodynamic form—much of it derived from extensive participation in racing—combine to give the Lotus an attractive specification. Below: Wide doors, with the window opening extending into the roof structure combined with body tumble-home, permit easy entry. The well-tailored seats, driving position and facia layout are suited to the character of this new car

There appears to be a very big market awaiting a successful British design in this class, to challenge the field which has been dominated by the Alfa Romeo Giulietta and the Porsche. The new Elite has the right specification, but at the moment it is only a prototype, and it will need to be proved. Colin Chapman is aware of this, and he intends to carry out a long development programme before the car is offered for sale; he knows that it may be necessary to make detail modifications as the Elite's development programme proceeds.

Initial production plans are for ten per week, but it will probably be twelve months before this target is reached.

SPECIFICATION

ENGINE: Coventry Climax Type FWE; No. of cylinders, 4 in line; Bore and stroke, 76.2 x 66.7 mm (3.0 x 2.625in); Displacement, 1,216 c.c. (74.23 cu in); Valve position, single o.h.c., wedge-type combustion chamber; Compression ratio, 10.0 to 1; Max. b.h.p. (nett), 71 at 6,700 r.p.m.; Max. b.m.e.p. (nett), 155 lb per sq in at 3,500 r.p.m.; Max. torque (nett), 76.5 lb ft at 3,500 r.p.m.; Carburettor, 1 S.U.; Fuel pump, S.U. electric; Tank capacity, 18 Imp. gal. (82 litres); Sump capacity, 8 pints; Oil filter, full flow, cloth element; Cooling system, Pump and thermostat; Battery, 12 volt 31 ampere hour.

TRANSMISSION: Clutch, single dry plate, 8in dia; Gear box, synchromesh on 2nd, 3rd and top. Central lever; Overall gear ratios, top 4.5; 3rd 5.94; 2nd 9.90; 1st 16.52 to 1; Final drive, hypoid 4.5 to 1.

CHASSIS: Brakes, Girling disc; Diameter of disc, 9½in; Suspension: front, independent with wishbones and anti-roll bar; rear, Independent, Chapman strut type; Dampers, Armstrong telescopic; Wheels, centre lock, wire spoke; Tyre size, 4.90—15in; Steering, rack and pinion; Steering wheel, 16in dia, 3 spoke; Turns, lock to lock, 2.

DIMENSIONS: Wheelbase, 7ft 4in (224 cm); Track: front and rear, 3ft 11in (119 cm); Overall length, 12ft (366 cm); Overall width, 4ft 10in (147 cm); Overall height, 3ft 10in (117 cm); Ground clearance, 6½in (17 cm); Turning circle, 31ft (10 m); Dry weight, 1,204lb—10¾ cwt (547 kg).

PERFORMANCE DATA: Top gear m.p.h. per 1,000 r.p.m., 16.7; Torque lb ft per cu in engine capacity, 1.03; Brake surface area swept by linings, 358 sq in; Weight distribution (dry weight), front 50 per cent, rear 50 per cent.

by Stephen Wilder - Jesse Alexander

a new strain of LOTUS

In 1955 Colin Chapman surprised everyone by getting his Hornsey firm, Lotus Engineering, elected to the SMM&T just in time to exhibit a Lotus chassis at Earls Court. The following year he stole the show by putting a Formula 2 race car on the Lotus stand, a car that had not even been test-driven, as the engine had no crankshaft and was connected to a dummy transmission. This year Colin has completed a minor hat-trick. By using the annual event to present the Lotus Elite, a Gran Turismo coupe of exquisite lines, he has not only established himself as a leading designer (see SCI, June 1957) but as a master of publicity and showmanship second to none in his field. The Lotus stand was again inundated, while the fancy limousines and exotic Continental sports cars were virtually neglected.

Typically, Chapman put the Elite on display before it had been driven and production is at least six months off. The first dozen or so off the line will probably be full-race GT cars with goodies straight out of the F-2 machines. With them, the Team Lotus will attempt to win a race-proved reputation.

The production Elite will have a 1220 cc Coventry Climax engine using the 1100's crank with the 1500's pistons. With a single S.U., output is 78 hp. Both cars will feature rack and pinion steering and the light-weight unequal wishbone i.f.s. with built-in anti-roll bar of the F-2 race car.

The cutaway drawing shows Chapman's newest idea in rear suspension, a fully independent arrangement of charming simplicity. The nearly-vertical spring-shock struts establish the attitude of the rear wheels by being firmly anchored to the hub assemblies. A wide trailing arm (to fix toe-in) combines with the doubly U-jointed-but-unsplined axle shafts to establish the arc of motion. Eliminating the sliding

Instruments include 8000 rpm tach, 140 mph speedo (!), ammeter and gauges for water temp and oil pressure. Adjustable bucket seats are well separated by tunnel.

A carefully shaped shadow-box neatly concludes the tail of this handsome coupe. Above the rear window appears to be provision for air vents to help cool the interior.

splines on the axles is advantageous because the friction involved in them when accelerating or braking over a bump creates a noticeable but irregular amount of friction damping in the rear suspension. On a larger car it is feasible to use recirculating balls on the splines, but on a car like the Lotus, the increase in semi-sprung weight would more than offset the benefits. The rear wheel geometry is rather like a swing axle set-up, but has much less camber change with deflection, which should be very good indeed. It's also possible to have lots of negative camber all the time without the ground clearance problem of a low pivot-point arrangement.

For the first time, Chapman is using fiberglas, and as might be expected, there are some novel touches. For one thing, it's a monocoque body, and for another, a fair amount of steel is bonded in where high strength and low bulk are necessary. A major member under each door sill provides beam strength. A rearward slanting tubular hoop serves as both roll-over bar and windshield frame. Attached to this hoop are the door hinges and a cross strut which carries the steering column, the hand brake, and the instrument panel.

Ahead of this hoop and above the forward extension of the door sills is a shear panel in the best aircraft tradition, but made of fiberglas. The firewall and rear bulkhead are also stressed members of the same material. Steel cross members carry the engine mounts and front suspension. A long fiberglas box, thoroughly separating the two bucket seats and surrounding the transmission and driveshaft, terminates at the bulkheads, desirably increasing torsional rigidity across the cockpit space.

Price? A bit over $4000 at ports of entry, says Jay Chamberlain, USA distributor for Lotus. Giulietta and Porsche drivers beware!

Rear parcel shelf is pretty well filled by the spare, even more so when optional second one is piled on top of the first. Tire size is 4.90 x 15 all around.

With the fuel tanks mounted in the front fenders and the 28 pound battery carefully tucked to one side, the full depth trunk is more than adequate for touring.

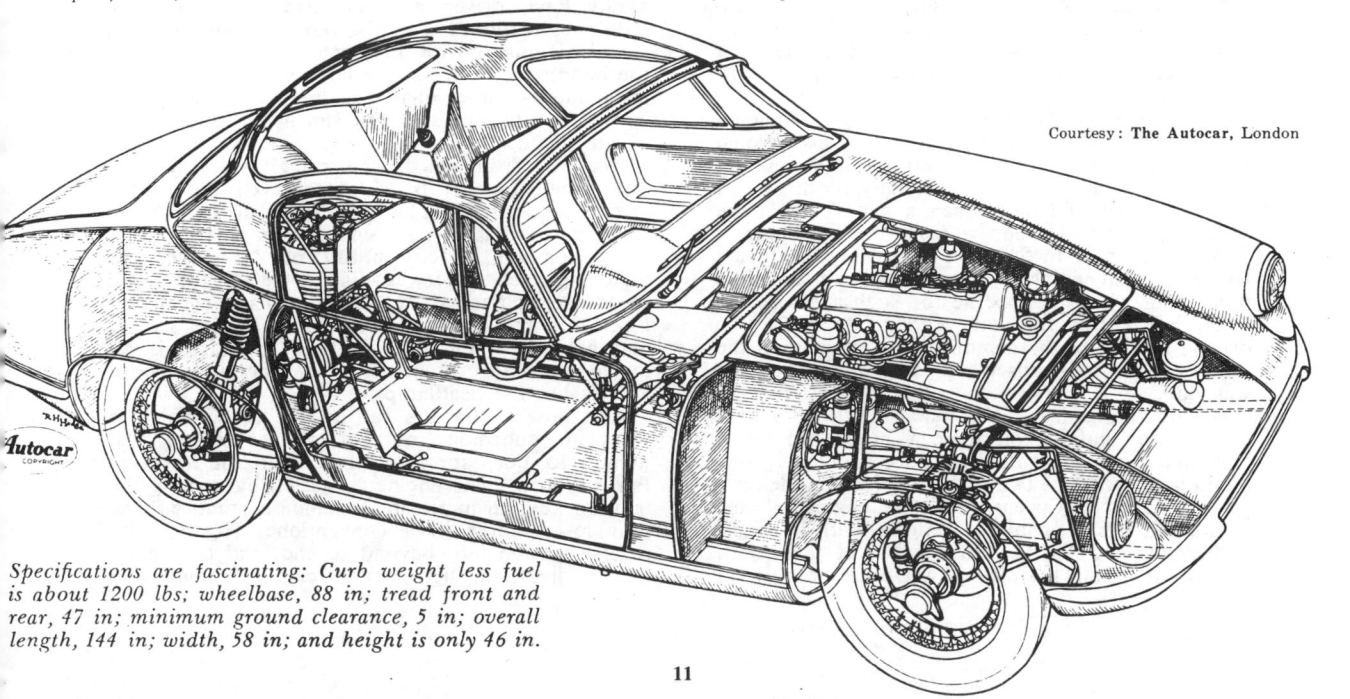

Courtesy: **The Autocar**, London

Specifications are fascinating: Curb weight less fuel is about 1200 lbs; wheelbase, 88 in; tread front and rear, 47 in; minimum ground clearance, 5 in; overall length, 144 in; width, 58 in; and height is only 46 in.

11

ROAD IMPRESSIONS OF
THE ELITE

6400 rpm in top and still accelerating . . . this little more than half a mile after turning on to the main Thame-Aylesbury road and less than two miles after making first acquaintance with the car outside the showrooms of Michael Christie Motors Ltd, at Haddenham. Foolhardy, some would say, but in a car like the Elite it is possible to feel " at home " so quickly—with all the controls situated, and operating, just as one would expect—that no real effort is required (on the driver's part) to reach and maintain speeds around the three figure mark.

In this connection it is immediately apparent that the level of wind noise is far lower in the Elite than in almost any other comparable car, while the exhaust note—which might be very much apparent to a bystander—is not sufficiently obtrusive to promote fatigue on a long journey although it is loud enough to give the driver a very good idea of engine rpm should he be too busy to glance at the tachometer.

In addition, the fantastic roadholding and phenomenal brakes of the Elite give it an exceptional safety factor, so that even if enthusiasm should get the better of wisdom there is usually room to sort things out with no more than a squeak from the tyres and a raised eyebrow from the passenger.

A run of four to five miles—possibly occupying less than that number of minutes—can do little more than whet the appetite for such a car as the Elite, but I was fortunate enough to share the driving on one of the demonstration models on the journey back to Hornsey and very illuminating it was, too.

But first things first. The Vale of Aylesbury on a crisp, sunny autumn morning, a bright red Lotus Elite and a stretch of road for once completely free of traffic. From the twin exhaust pipes the delightful burble of a Coventry Climax engine. Some would call this a " cooking " engine, with its single carburetter and Stage 1 camshaft, but it is both free-revving and flexible, powerful—especially in respect of such a light, sleek car—and, it would appear, economical.

From generalities to details. Despite the very low build of the Elite it is no more difficult to get into and out of than other more conventional sports coupés. The seat backs are comparatively flat, making it easy to slide into the car, but on the move they provide a surprising amount of lateral support—evidence of the thought given to their construction.

Unlikely as it may appear from the above photograph, I find the legroom of the Elite quite adequate, and although there is little room between the steering wheel and my knees I experienced no discomfort during a journey of 50 miles and would gladly have driven a much greater distance. (I didn't really notice it at the time, but afterwards I found a bruise on my knee where it had rested on the handbrake.) When the left foot is not in use it rests under—rather than beside—the clutch pedal. There is sufficient seat adjustment to suit drivers of all heights.

The view from the driving seat gives a very good indication of what is to come, even before the engine is started. The wood-rimmed light alloy steering wheel is in the " enthusiast's dream " category, and set directly ahead of the driver are matching tachometer—reading to 8000 rpm—and speedometer, flanked by fuel contents gauge, combined oil pressure and water temperature gauges and ammeter. On the right of the facia is an ingenious " trigger " switch which sounds the horn and flashes the headlamps.

And now to try the car on the road, which is what an enormous number of people are just itching to do, especially after seeing it again at Earl's Court. The engine is started by turning the ignition key, an increasingly common practice. The gear lever works in a conventional H-gate, with reverse on the left, beyond second, and the action of the hydraulically-operated clutch is light and positive. The throttle-linkage seems well contrived and the car moves off smoothly and deceptively quickly from rest, application of a heavy right foot swinging the tachometer needle very rapidly round its dial in first and second gears.

Third is engaged on the approach to a gentle bend but there is no need to lift off and the Elite goes round in a most demure fashion, with hardly any movement of the steering wheel. Into top and the road unfolds ahead, the sloping bonnet providing excellent visibility and the upstanding wings assisting in accurate placing of the car for corners and mobile hazards.

Two more gentle curves, first to the right and then to the left, and there is the main road; a touch on the brakes, sweetly down into third, not so sweetly down into second (wide ratio gears) and very slowly round what amounts to a hairpin bend with an unusually small amount of steering lock.

ROADHOLDING

This is one of the things that makes the Elite such a joy; light, accurate steering, along with other simple positive controls, makes for precise driving, and it is really far easier to handle a car of this type than most family saloons. In fact, the manner in which the car achieves what are, to most people, very high speeds demands much greater concentration and awareness of responsibility from the driver. Nevertheless, anyone accustomed to high performance cars should delight in the responsive handling of the Elite. The roadholding is in the racing car category in that the tail can be broken away quite easily without any danger of a complete loss of adhesion, and steering with the throttle materially increases potential cornering speeds. In all this the special Firestone tyres obviously play a very important part.

But back to A418. A long right-hand bend and then a gentle right and left are treated as if the road were quite straight, the car feeling completely stable and showing very little sign of roll. On to the straight, over the railway bridge, up to 6000 rpm—6200, 6400 in top and then reason says "that's enough" at such an early stage in an acquaintance with a very new motor car. And anyway there are corners ahead, right, left, right and then immediately right again on to the by-road. Back into the village at a more respectful speed, down into second behind a large agricultural contrivance occupying most of the street ("what be these secret new cars they keep bringing through here") and that's almost the end of the run. Just time to try the brakes ... up to seventy ... brakes hard on ... and down to twenty in the time it takes to read this sentence. Some deceleration!

That seems to be that. Time for no more than brief impressions, impressions of acceleration, braking, handling, interior layout—impressions perhaps a little confused but nevertheless extremely favourable. But this is not all. The little red projectile has to go back to Hornsey, so a quiet word with Mike Costin and all is arranged. Before departure yet another impression. A considerable amount of luggage, such as would embarrass the boot space of one or two current sports cars, is stowed away with ease; the shape of the tail is as practical in this respect as is that of the rest of the car from the viewpoint of functional efficiency. There is also a very useful amount of stowage space inside the car, both in the deep door pockets and parcel tray and on top of the nicely-trimmed spare wheel cover.

Earlier ideas on the Elite are confirmed on the Watford By-Pass. The miles just melt away and the hindrance of traffic is reduced to a minimum. The acceleration which is available reduces overtaking time to a minimum, while the feeling of security imparted by the brakes permits a very rapid approach to roundabouts and similar obstructions. And when, finally, it seems prudent in urban surroundings to travel with the stream, the Elite potters along without any sign of temperament—a truly versatile motor car.

POSTSCRIPT

After writing the above I was fortunate enough to have the opportunity of driving another Elite—a blue one this time—after dark, and thus to get an idea of its potentialities for a long journey when the roads are clear (at the time they weren't), or just for going to the cinema. (I know; if you had an Elite you wouldn't want to go to the cinema.)

Not far from Earls Court the occupants of a large American car beamed down and suggested "swap?"—I declined. Gradually losing some of the traffic I found time to play with the lights switch, which pulls out to operate side/tail—and head-lamps and turns to control an interior lamp, mounted above the rear window, and the rheostatic instrument lighting. The main beams were set a little low for really fast driving (I was taking the car back to Hornsey for the setting to be adjusted) but provided a very adequate spread.

In general, my earlier impressions of the car were all confirmed on this evening trip from Earls Court to Hornsey via some of the nearest countryside to London.

CONTROLLABILITY

This car was fitted with Michelin "X" tyres, which provided absolutely leech-like roadholding, to the extent that I felt the tail would never break away in the dry, even in second gear. There must be a limit to adhesion, of course, but my own opinion—and this is a purely personal assessment—is that I wouldn't like to find it in a tight corner. For the driver who prefers to corner geometrically the combination of Lotus independent suspension and Michelin "X" tyres must be just about the ultimate, and in this respect the Elite sets a standard for roadholding which few other road cars can equal (the AC Ace is the only one which comes immediately to mind). But for controllability, for ample warning of breakaway and for what might be called generally more responsive handling I felt happier in the Firestone-shod car. Only a much more extensive comparative test, in both wet and dry conditions, could finally decide the issue, of course.

Finally to an important point which is often overlooked in assessments of high performance cars—fuel consumption. The Elite travelled along a well-known section of road at much the same speed as a number of other thoroughbred sports cars I have recently tried; the only difference is that it was covering about 35 miles per gallon of petrol, whereas the others were doing rather less than 20.

Now that I have tried the car in town and country, in morning sunshine and evening mist, my tongue is just hanging out in anticipation of a really long run in the Elite.

PPS. Colin Chapman says my tongue is likely to get awfully dry.   D.P.

ON TH

Aerodynamic excellence coupled with good looks and practical features—that is the Lotus Elite. The bonnet falls away sharply for good vision but the wings are prominent and greatly aid high-speed judgment. Profile would be even smoother with acetate covers for headlamps

COLIN CHAPMAN'S Lotus Elite undoubtedly stole the 1957 Motor Show at Earls Court. Its beautiful proportions, obvious performance potential, and novel "plastic" construction provided motoring journalists with more material for copy than they had had for a long time. Since then two prototype cars in the hands of Ian Walker and Tom Lawry have been engaged in British motor races and have collected a huge number of race successes, a reputation for near-invincibility, and a great deal of development data.

Now, twelve months after its introduction the Elite is in full production. Although externally identical with last year's Show car numerous structural and mechanical modifications have been carried out in the light of experience. The engine has been mounted further back in the "frame", and has been raised half an inch to improve ground clearance and to provide more space for the silencers. A thermostatically-controlled electric cooling fan is now fitted behind the radiator, where it cuts in when the water temperat builds up.

Several structural changes have been made. Experie has proved the metal reinforcement at the final-drive rear suspension mounting points to be superfluous, and these highly stressed areas are now of increased thickn and carry rubber bushes. The basic structure now c prises two skins, back to back, the inner one being produ on male moulds so that the interior and external finish of equally high standards. This is one of the most strik features of the latest Elite, even the bonnet and boot interiors being smooth and polished. It will undoubte prove to be a major step in reducing buyer resistance glass-fibre construction, and its normal "shredded whe internal finish.

In addition the moulds have been modified to reduce number of joints, so that body loads have fewer joint p to cross. For the time being Elite production will be at

The single o.h.c. Coventry Climax engine has been mounted further back and slightly higher for 1959. Note single S.U. carburetter and excellent under-bonnet panel finish

Boot is huge for a car of this type, and is trimmed with carpeting. Battery (12 volt) is centrally mounted and par of the spare wheel "bulge" projects into boot

ROAD WITH THE LOTUS ELITE

r week rising early in the new year to ten per week when
e Lotus organization moves to its new factory. At the
ne of this brief trial nine Elites, including Walker's and
wry's cars had been built.

On a recent short test of the latest Lotus Elite from
tus U.K. Distributors, Alexander Engineering (Sales)
d of Haddenham, Bucks, *S.C.I.* was once again greatly
pressed with the beautiful, wind-cheating lines of the
te, and with its many practical features for the motorist
o likes to go far and fast.

One of the most amazing features of the Elite—in view of
staggering performance is the retention of a single 1½-in.
U. carburetter, a cast exhaust manifold of a rather sinuous
pe, and a Stage I camshaft to its 1,220 c.c. Coventry
max single o.h.c. engine. A change from last year is the
e of the light-alloy cylinder block in place of the original
t-iron type.

With a power-output of 75 b.h.p. at 6,100 r.p.m. the
rformance is staggering, yet the car is flexible and will
tter along on a 4.5 : 1 top gear without a snatch. On that
ne top gear it will snarl up to 115 m.p.h. with sports/racing
acceleration, but there is always that wonderful feeling
security afforded by the 9½-in. Girling disc brakes. On
e car we drove the brake pads were of competition material
ich promoted a certain amount of "nose-wag" when
y were applied from high-speed, but at no time did they
ove embarrassing. Production models will be fitted as
ndard with a "softer" lining which makes for "all-square"
pping. The competition linings also promote a roughness
ich can be felt inside the car when braking from low
eeds, but the production models will have this fault
dicated.

Colin Chapman explained that by the fitting of twin
burétters, a racing type exhaust manifold and a Stage II
nshaft, another 25 b.h.p. can be obtained! This should
se the top speed to around 125 m.p.h., and the already
pressive acceleration figures.

Driving the car is indeed a wonderful experience enhanced
the tailor-made feel. The perfectly-shaped bucket seats
upholstered in soft hide and provide first-rate support
der the fastest cornering manoeuvres. The windscreen
is in dull black glass-fibre, and as well as giving the
pression of a black leather finish successfully combats
y unwanted reflections in the curved windscreen.

The Editor deposits Firestone rubber on the Buckinghamshire lanes as he demonstrates the Elite's impressive acceleration. From any angle the car is unusually handsome

The driving position is perfect for fast travel, the seats being adjustable for the longest-legged driver, and the small diameter wood steering wheel falls nicely into the lap. The instruments are ahead of the driver, mounted on a raised "binnacle", and consist of 4-in. dial Smiths rev counter and speedometer, and small diameter oil pressure and water temperature gauges. All have black dials and rims. Starting is effected by turning the ignition key to the right in the modern manner (centre of facia) and the choke, trafficators (time-switch), lights and other controls are neatly grouped to the left of the starter key.

The M.G.A. gear-lever projects from the large gearbox/prop-shaft tunnel (a contributing factor to the rigidity of the plastic *monocoque* structure) where it is positioned ideally for lightning changes when exploiting the very impressive performance. Behind the two seats is the spare wheel, neatly covered with high-quality carpeting, and leaving a reasonable amount of space for luggage. The tops of the two rear suspension units project into the rear compartment.

The boot contains only the small 12-volt battery and for

Continued on page 71

High-speed luxury. Seats and trimming of the Elite are of the highest standard. There are door pockets, a parcel shelf, and an ash tray in the propellor shaft tunnel

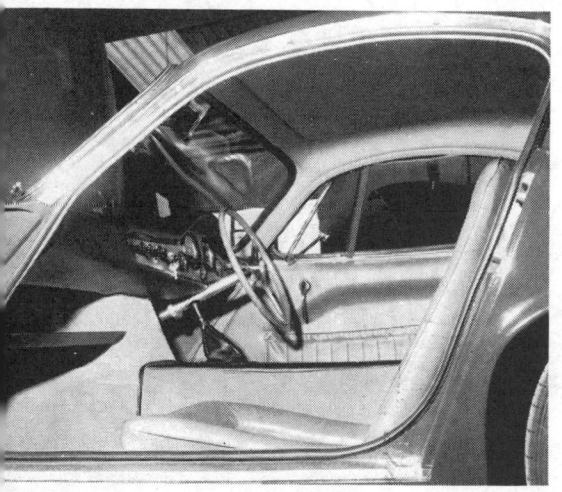

Spare wheel is carpet-covered and sits behind the seats still leaving space for a small suitcase and parcels. Suspension units project into the interior. Rear window is large

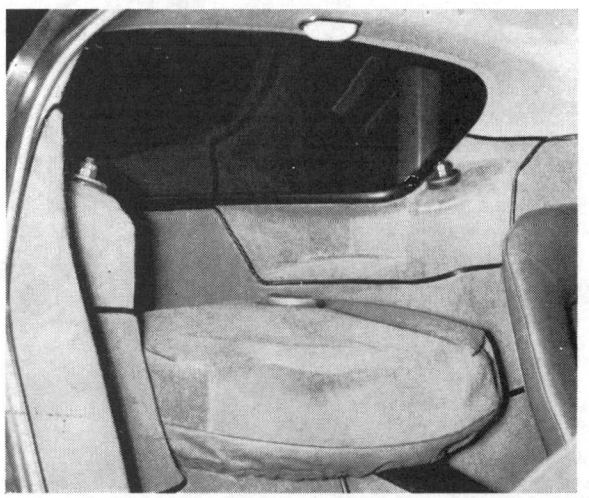

FIRST TEST RUN OF THE
LOTUS
ELITE...

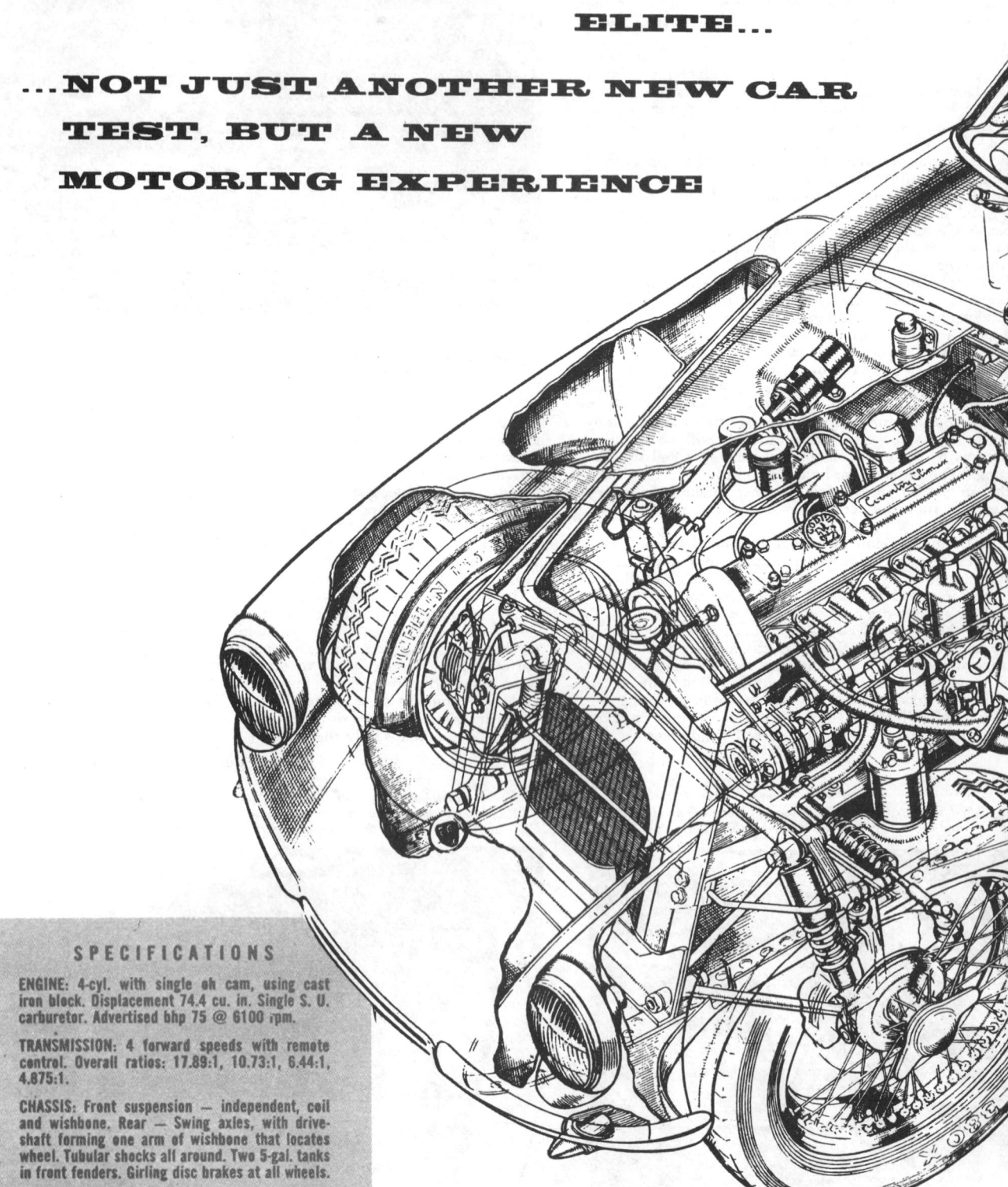

...NOT JUST ANOTHER NEW CAR TEST, BUT A NEW MOTORING EXPERIENCE

SPECIFICATIONS

ENGINE: 4-cyl. with single oh cam, using cast iron block. Displacement 74.4 cu. in. Single S. U. carburetor. Advertised bhp 75 @ 6100 rpm.

TRANSMISSION: 4 forward speeds with remote control. Overall ratios: 17.89:1, 10.73:1, 6.44:1, 4.875:1.

CHASSIS: Front suspension — independent, coil and wishbone. Rear — Swing axles, with driveshaft forming one arm of wishbone that locates wheel. Tubular shocks all around. Two 5-gal. tanks in front fenders. Girling disc brakes at all wheels.

DIMENSIONS: Wheelbase 88 in., overall length 144 in., overall height 46 in., overall width 58 in., front and rear tread 47 in. Weight 1205 lbs.

Illustration by
James A. Allington

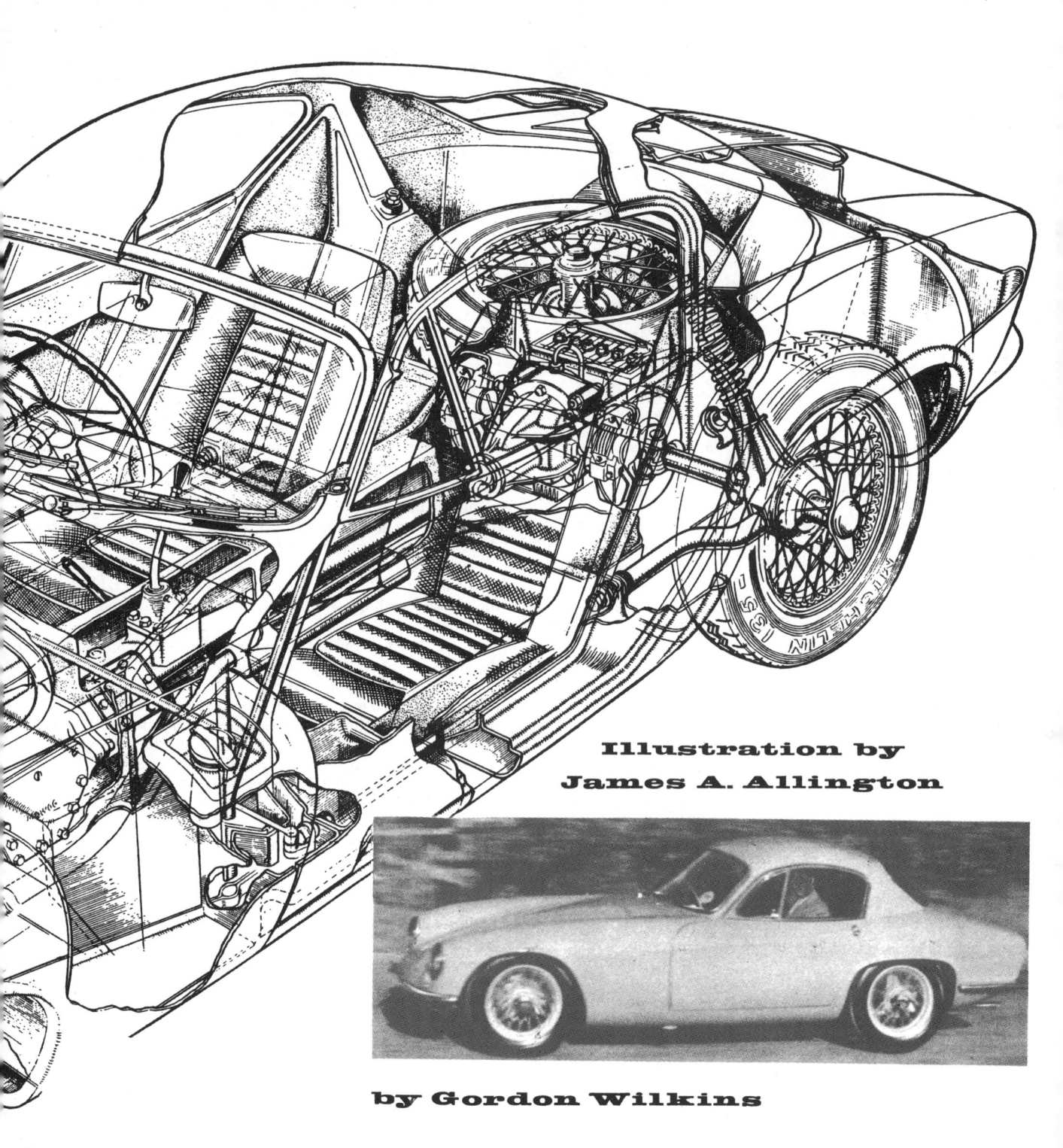

by Gordon Wilkins

IT WAS A WARM SUNNY MORNING in fall. The leaves were turning to gold on the trees and the smoke from the farmhouse chimneys rose straight up in the still air. The quiet of the English countryside was suddenly assaulted by a crisp snarl rising and falling, sometimes cutting out for a moment, then rising almost to a scream. Soon there was a glimpse of red moving very fast and a tiny, brilliantly colored car hurtled through the twists and turns of the winding country roads.

It was my first test run on a production-type Lotus Elite, but to me it was more than just another new car test — it was a new motoring experience. This revolutionary little coupe, with its all-plastic structure, all-independent suspension and high power-to-weight ratio brings racing car roadholding into the Gran Turismo class. It holds 100 mph with insolent ease and has the inbuilt safety margins which only competition-bred cars can provide.

Since the first hastily finished prototype was displayed at the London Motor Show in October, 1957, a whole year's development work has been done on the race tracks (where it won nine records and set up five new lap records for its class), on the road and at the M.I.R.A. proving ground. The structure has been refined and simplified and production models are now coming through at the rate of two a week, which should rise to 10 a week in 1959.

The Elite looks tiny as you stand beside its 46-inch-high roof, on which you can rest your elbows. Getting in over the deep body sill requires a bit of agility. Once in, there is stretching room for six footers with leg-, head- and elbow-room to match.

With me on this test run was a six-foot-four colleague who settled down quite happily. Pedals are perfectly arranged, with a long spoon-shaped accelerator positioned for heel-

LOTUS

Seats show vast racing experience, have firm cushions with softly rolled edges. Small wheel is well away from chest.

continued

and-toework when making swift downward shifts. To the left of the clutch are two buttons: one to dip the headlights and the other to work the windshield washer. I noted carefully which was which, so that I would not suddenly be confronted with a wet windshield when facing dazzling headlights.

The seats show the benefit of vast racing experience. The center of the cushion and backrest are firm. Around the edges are softer rolls which locate the body laterally on fast corners without appearing to insist. The backrest rises almost neck-high and invites the passenger to relax. The wheel, with wooden rim and light alloy spokes, is small and well away from the driver's chest, encouraging the straight-arm type of driving position, permitting even sharp corners to be taken in one clean sweep. Steering is quick enough to cope with any situation, for it needs only 2½ turns lock-to-lock. Once the Elite really gets going you are rarely conscious of moving the wheel at all, for this is a car you steer with the wrists — and with the throttle.

The artistic asymmetrical instrument panel has a 140-mph speedometer, tachometer with red warning sector from 6500 to 8000 rpm, and a group of fuel gauge, ammeter, water thermometer, and oil pressure gauge. Lined up to one side are knobs for choke, heater fan, wiper, light switch, ignition-starter switch and time switch for direction indicators. On the other side, where the fingertips can reach it without the hand leaving the wheel, is a trigger for horns or headlights.

Tiny Elite, only 46 inches high, is somewhat difficult to enter. Once inside though, there is legroom for six-footers.

The short gear lever lies on the deep center tunnel just where the hand naturally reaches for it. Lower down is the handbrake.

The spare wheel lies flat just behind the seats, under a neat carpet cover, leaving space for a second spare on top or for small luggage and parcels. Fuel tanks holding 8.4 gallons are in the front fenders, leaving the whole tail free for a really useful luggage trunk capable of holding several big suitcases. Without luggage, all major masses are concentrated within the wheelbase.

The final look around shows excellent vision over the low-swept hood and through the wide curving rear window, so now let's go. As the engine bursts into song it becomes apparent that this particular car has the competition silencing system. More extensive decibel suppression will be supplied for those owners who want their cars mainly for road use. Clutch action is quite light, and engagement is not fierce, but it goes in with a firm no-nonsense bite. As you tramp hard on the loud pedal ready to ease back at the first sign of wheel-spin, you get the first surprise. This independent Lotus rear suspension takes all the torque there is with no trace of spin, and the rev counter needle goes swinging around as the car shoots forward with both wheels biting. At 4500 rpm the engine is beginning to scream and it comes as a surprise to find there are still 2000 rpm in hand. When you use the lot there is real action. With the 4.55 to 1 final drive in the car I was trying, 6500 rpm gave 29 mph in first, 49 in second, 82 in third and 109 in top. Using 5000 as a comfortable limit, the upward rush brings 23 mph in first, 37 in second, and 63 in third.

During my first mile at the wheel I was twitching the Elite through an S-bend at 80, then braking with a quick downshift to second, then first for a hairpin bend onto a main road where it went up to 100 mph without pausing for breath. The plastic structure feels as taut as a drum, there is no perceptible roll even on sharp bends taken at the limit of adhesion, and the response of the controls is so beautifully graded that the car seems to sense exactly what is expected of it.

After a few miles the Elite had imparted so much confidence that it seemed quite natural to be cornering with the throttle, helping the tail around or checking it almost without conscious effort. The car I tried had special Firestone tires with nylon carcass, which seemed to encourage a fluent style of driving.

Engine noise in the trim in which I tried it was obtrusive enough to become tiring on a long run, but wind noise seemed to be very low. With the touring-type silencing, everything should be right for covering long distances in comfort at awe-inspiring average speeds. The ride is firm over bumpy sections, but smooth and steady when cruising at 80-90 mph on main roads.

A quick speedometer check showed a correction of four mph at 60, so with due allowance for this the watches were produced for a standing start acceleration. First time, from 0 to 60 mph took 12 seconds. Next time, with better coordination of throttle, clutch and gear lever, it was down to 11.5 seconds. There was no time for more, but 10-11 seconds should be possible. As for stopping power, with four-wheel disc brakes (inboard at the rear) it is almost superfluous to say it was superb. It is not simply a question of having good brakes; the all-independent suspension really keeps the wheels on the ground, pulling the car up all square.

On this early example there were various details which still needed attention; the fit of the doors could be improved, and shock absorber settings were still the subject of experiment. Wisely, Colin Chapman is delivering the first 100 cars to buyers in Britain, so that he will be able to watch their behavior. It is therefore likely to be several months yet before the first deliveries are made to the U.S. but it is something well worth waiting for.

road test

LOTUS ELITE

Cozy, comfortable and quick, for the well-heeled enthusiast

EVERY SO OFTEN a car comes along that represents the ultimate state of the art in some respect—performance, appearance or utility—and it immediately captures the fancy of some segment of the auto buying public.

The Lotus Elite is such a car. From the first time the Elite was shown—at the London Auto Show at Earls Court, in 1957—sports car enthusiasts have been eager to get their hands on one of these little gems.

With the importer—Jay Chamberlain—located in North Hollywood, Calif., just a few miles from our offices, it seemed to be an easy matter to get an Elite for a road test. Jay brought two cars to California after Sebring last March, but neither was available for a road test. The car's designer and builder, Colin Chapman, had decreed that *no* Elites were to be given out for road tests. Driving impressions, yes; road tests, no.

A driving compartment designed for maximum control.

Surprising flexibility is obtained from Climax engine.

PHOTOS BY POOLE

19

Our only possibility, then, lay in getting one from a private owner. The Elite ultimately tested was No. 1047, owned by Bill MacDonald, of Beverly Hills, Calif. Bill very kindly consented to let us test his newly acquired jewel and, while the owner drove the car during actual acceleration and top speed runs, we drove the car for our own driving impressions (augmented by driving impressions in Chamberlain's demonstrator).

The first impression of the car, and one that is not apparent from the photos, is of its extremely small dimensions. Standing beside the car it would seem all but impossible to squeeze in a man of even moderate size, let alone one six feet or more tall. First surprise: One of our test crew is six feet tall and another is six feet two, yet neither had any difficulty with either leg or head room. The steering column is not adjustable, but the seats have a fore and aft adjustment that can compensate for the leg length of any driver between five feet and about six-three. However, a driver or passenger of more than average girth might find things a little snug about the posterior, because of the form-fitting bucket seats.

The consensus of all who drove, or rode in, the Elite was that these seats are among the very best we've ever tried, for comfort. The almost vertical, wood-rimmed, spring-type steering wheel is placed in a manner best suited to the straight-arm driver.

The moderately instrumented panel is functional in the truest sense and contains speedo, tach, water temperature, oil pressure, ammeter and fuel level gauges. A rheostat controls the lighting of the circular instruments (white numbers on black) for night driving and we found that even with the instrument lights on at their brightest the light was not obtrusive—in fact the instruments proved just a little difficult to read at night because the numbers were so small. The cowled panel prevents any windshield reflection from the instrument lights and the mat finish of the inside cowl assured no reflections in the windshield during daytime driving.

Although the interior appointments are adequate, and generally well executed, the window glass (which is removed and placed in a container, then stored behind the seats or in the trunk) is a little inconvenient. Locking the car, for example, would mean digging the windows out of wherever they are stored and inserting them before leaving.

Our first ride in an Elite was disappointing. Maybe we had expected too much. It was so firmly sprung as to be quite stiff (even by the oft-referred-to sports car stand-

Not exactly cavernous, but adequate. Battery at left.

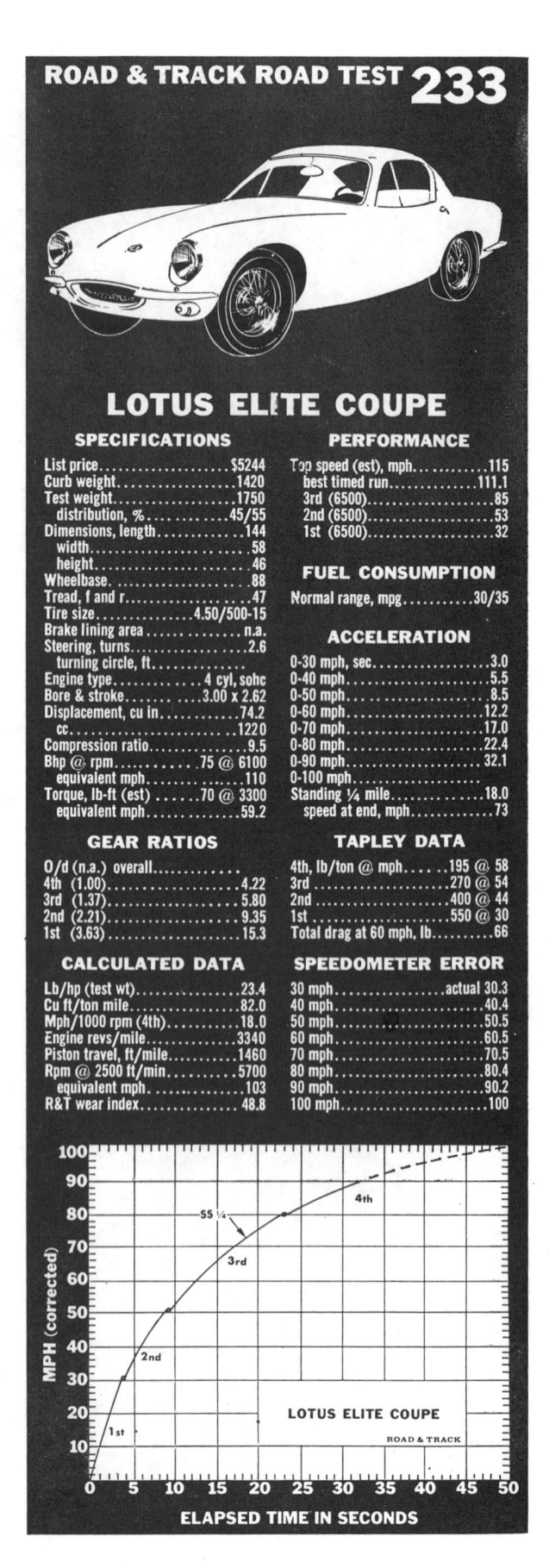

ROAD & TRACK ROAD TEST 233

LOTUS ELITE COUPE

SPECIFICATIONS

List price	$5244
Curb weight	1420
Test weight	1750
distribution, %	45/55
Dimensions, length	144
width	58
height	46
Wheelbase	88
Tread, f and r	47
Tire size	4.50/500-15
Brake lining area	n.a.
Steering, turns	2.6
turning circle, ft	
Engine type	4 cyl, sohc
Bore & stroke	3.00 x 2.62
Displacement, cu in	74.2
cc	1220
Compression ratio	9.5
Bhp @ rpm	75 @ 6100
equivalent mph	110
Torque, lb-ft (est)	70 @ 3300
equivalent mph	59.2

GEAR RATIOS

O/d (n.a.) overall	
4th (1.00)	4.22
3rd (1.37)	5.80
2nd (2.21)	9.35
1st (3.63)	15.3

CALCULATED DATA

Lb/hp (test wt)	23.4
Cu ft/ton mile	82.0
Mph/1000 rpm (4th)	18.0
Engine revs/mile	3340
Piston travel, ft/mile	1460
Rpm @ 2500 ft/min	5700
equivalent mph	103
R&T wear index	48.8

PERFORMANCE

Top speed (est), mph	115
best timed run	111.1
3rd (6500)	85
2nd (6500)	53
1st (6500)	32

FUEL CONSUMPTION

Normal range, mpg	30/35

ACCELERATION

0-30 mph, sec.	3.0
0-40 mph	5.5
0-50 mph	8.5
0-60 mph	12.2
0-70 mph	17.0
0-80 mph	22.4
0-90 mph	32.1
0-100 mph	
Standing ¼ mile	18.0
speed at end, mph	73

TAPLEY DATA

4th, lb/ton @ mph	195 @ 58
3rd	270 @ 54
2nd	400 @ 44
1st	550 @ 30
Total drag at 60 mph, lb	66

SPEEDOMETER ERROR

30 mph	actual 30.3
40 mph	40.4
50 mph	50.5
60 mph	60.5
70 mph	70.5
80 mph	80.4
90 mph	90.2
100 mph	100

Beauty is as beauty does, but it's poorly protected.

ards) and the noise level was higher than we expected.

This first ride, as we later found out, was quite misleading and not representative of the production models. Chamberlain had taken us out, with strong protests on his part, in his ex-Sebring competition car which had two-stage shock dampers set on their stiffest setting, tires carrying 30 psi (on a 1420-lb car), a steel timing gear in the engine (which made it sound a little like a coffee grinder) and little or no attention to soundproofing in the car.

The "street" version of the Elite is a different car entirely. It's still sure-footed, with the comfortable seating and responsive steering of the competition car, but the noise level is as low as in any touring car and the ride is so smooth as to belie the ultra-light weight.

A right hand drive car, as all Elites are, is a little inconvenient for U.S. driving but, surprisingly enough, we found one of the worst blind spots to be over the right shoulder, toward the rear, rather than to the left rear. A rear-view mirror on each front fender is highly recommended, even though it slightly spoils the aesthetic appeal of the car.

Shifting left handed presented no problem either, after the first few miles behind the wheel, but it is certain that your self-winding wrist watch won't run down from lack of motion.

The full independent suspension—wishbones in front and Chapman-designed strut and trailing arms in the rear (both using combined coil spring/damper units)—gives a rather odd feeling on rough or undulating surfaces. This is especially noticeable to drivers who are relatively unfamiliar with 4-wheel independently-sprung cars.

As a consequence, a skittish feeling persisted until we became used to the Elite, and the longer we drove it the more natural, and likable, things became.

Corners taken at what seemed maximum or near maximum speed made us feel a little foolish after completion because they had turned out to be so easy. Invariably we felt that we could have taken the corner much faster than we did. Cornering can be done as flatly as with any car we've driven and in spite of a curious "walking on its toes" sensation the Elite followed the road as though it had eyes of its own.

Acceleration is not spectacular, as can be seen from the data panel, but is good and the car seems to be at its best at higher speeds. Drivers we've talked to who have had experience in other Lotuses (Mark VIII, IX or XI) report that the Elite doesn't have the heavy understeering that is so pronounced in the race cars. At the speeds we drove the Elite it displayed almost dead-neutral steering and on a few occasions even over-steered. The hydraulically-operated Girling disc brakes required a little more than ordinary pedal pressure, but worked like a charm. Not the emergency brake, however. It barely worked at all, but adjustment should take care of that.

The hydraulically operated clutch, a single dry plate, 8 inches in diameter, transmitted the power in a rather sudden manner, but always positive, never slipping or chattering, regardless of the driver's actions.

To sum things up, this is a car for enthusiasts only. And, for the enthusiast, it represents a car that "looks right," embodying an excellent body design from both the aesthetic and functional standpoints, good performance for its class, superb handling and comfortable seating. On the debit side, the only real criticisms are the poor parking protection, common to many sports cars, and the high initial cost.

The potential buyer can only balance his desire for the car against his ability to pay for it, and then make his decision. Its unique fiberglass unit body/frame construction is the only example of its kind in the world, and one thing is certain: an Elite owner won't find too many other cars like his on the road.

The Kamm theory helps keep the rear end to a reasonable length on the aerodynamically clean body.

ROAD TEST: LOTUS ELITE

Lotus Elites had an enviable competition record in 1959, and one of the most successful cars in European events was Peter Lumsden's which won the 1300 cc G.T. class in the Nürburgring 1000 Km. race and the 1500 cc class at Le Mans — all during the owner's annual vacation. In the 24-Hour race the dark-green Elite covered nearly 2300 miles, and as well as winning the 1500 cc category against strong sports-racing opposition it finished eighth overall and second in the *Coupe a l'Indice au Rendement Energetique* — a classification based largely on fuel consumption. What better car to try as an example of an Elite modified for racing?

Regular SCI readers will be aware that the Lotus Elite is unique among European cars in having an all-fiberglass chassis/body structure, containing very little in the way of metal other than engine and suspension mountings and a tubular hoop around the windshield. To this shell are attached the standardized Lotus wishbone and coil spring front suspension and the independent "Chapman strut" rear end. Power unit is the well-known Coventry Climax single-overhead-camshaft engine, enlarged to 1216 cc and designated FWE, mated to an MGA gearbox.

Peter Lumsden's car was a pre-production prototype which did a lot of development work before being stripped, rebuilt and sold to him at the start of the 1959 season. For this reason the sceptics regarded his visit to the bumpy, chassis-breaking Nürburgring as a waste of time. But the car emerged from this ordeal, and the less exacting day-and-night at Le Mans, with colors flying, although its differential housing had come adrift a few weeks earlier during a club meeting at Silverstone! In such ways are faults discovered.

Another aspect of the car's prototype nature is the bonking transmitted through the rear suspension units and the general level of noise which results from an absence of sound insulation. In addition the exhaust note, with the small racing muffler fitted, becomes very "sporting" above 4000 rpm, and also gives rise to resonances inside the car which might be less tolerable on a touring run than on the race track. On current production cars the bonks have been eliminated and the noise level has been considerably reduced, but a certain amount of resonance remains, as does gearbox and transmission noise. Of course, some people regard these items as essential features of sports car motoring.

Even in standard form, of course, with a single carburetor engine, the Elite is no mean performer and probably has better handling than any other small closed car — and a lot of open ones too. Its Girling disc brakes stop it from speed in incredibly short distances and its light, positive controls make it enjoyable to drive under all conditions.

In view of all this, Elites require far less modification to turn them into successful racing machines than the majority of production cars. The standard car behaves so well in the wet, and even on snow, that the chassis can obviously take more power without any snags arising. And these are exactly the lines on which Peter Lumsden worked when he went about having his Elite modified for use as a "racer."

The chief, and almost only, departure from standard on Lumsden's is the fitting of a Stage III cylinder head (see page 38 of this issue), with its five-bearing camshaft, together with twin 1½-inch SU carburetors on special intake manifolds and a four-branch exhaust manifold. Compression ratio is 10.5 to 1 and maximum power is raised from 75 bhp, at 6100 rpm to 102 at 7000. For long-distance events an oil cooler was fitted, together with baffles around the rear brake discs to direct heat away from the interior of the car. To combat under-hood heat a small aluminum duct was fitted to provide cool air for the carburetors. In character with the "racing" engine, close-ratio gears were fitted; surprisingly enough these proved to be quieter than the standard gears.

One of the secrets of Peter Lumsden's success was the car's complete reliability. At the Nürburgring the Elite ran faultlessly, building up an early lead in its class and increasing it as the Alfa-dominated entry was forced to stop for tire changes and brake adjustment. On some sections it proved — due to its excellent handling — to be as fast as the leading Aston, and faster than some of the Ferraris. Peter and his co-driver, Peter Riley, drove to a set lap time, and this proved to be high enough for them to win the class by eight minutes.

The same approach to long-distance racing paid off at Le Mans, where no mechanical work of any description was necessary throughout the twenty-four hours. During each 32-lap session the Elite used 15 to 17 gallons of gas and two to three pints of oil. On Sunday morning the radiator and gearbox were topped-up — just for good measure — and that was all. The target speed was maintained throughout the major part of the race, and toward the end it was possible to slow down considerably and still maintain the class lead.

Although the Elite is so low-built, the big, wide-opening doors simplify getting in and out, and once inside there is ample head- and leg-room, even for my 6 feet 4 inches. There is also a surprising amount of luggage space, both in the deep rear trunk and inside the car, where the spare wheel is carried — out of sight — on a ledge beneath the rear window. Trunk and hood lids have over-center hinges and stay open without supports, and hood-opening provides easy access to most of the items requiring regular attention.

The driving position is fairly upright, with the seat well away from the very pleasant, wood-rimmed steering wheel. The seats are comfortable but don't provide a great deal of lateral support for people of slender build, and although the deep propeller shaft tunnel helps in this respect, Peter Lumsden has had small padded side-pieces fitted at the junction of the seat cushion and back-rest.

The control and instrument layout is almost ideal as standard, and thus Peter Lumsden has seen no reason to change it. The pedals are well spaced, with ample room for the left foot below the clutch. The accelerator extends downwards to facilitate heel-and-toeing. Below and to the left of the clutch is a foot-operated dimmer switch, with a windshield-washer plunger alongside it.

The short gear-lever protrudes from the transmission tunnel within a few inches of the steering wheel, and the gearshift — although rather stiff initially — has loosened up considerably as a result of its racing use.

Directly ahead of the driver is the comprehensive instrument panel, with matching 8000 rpm tachometer and 140 mph speedometer flanked by a fuel gauge on the left, a

Peter Lumsden's Lotus Elite is taken through the Esses during the running of the 1959 Le Mans race. Its happy owner came home first in class.

conbined oil pressure/water temperature gauge in the center and an ammeter on the right. Pull-out switches control choke, two-speed wipers and lights (parking, head and interior lamps all on one switch), and a toggle switch on the right operates the horn and flashes the headlamps — a useful item rarely found on British cars.

On the road, one of the most remarkable things about the car is the lightness of the rack and pinion steering, and, in fact, all the other controls. It's rarely necessary to move the steering wheel rim more than a few inches, and even in the wet the slightest movement of the wheel serves to correct an incipient slide. Anyone accustomed to low-geared "spoon-and-jelly" steering might find the instant reactions of the Elite a little disconcerting at first, but after a few miles it becomes almost second nature to guide the car swiftly and smoothly through bends, and to hold the wheel lightly between the finger-tips on the straight.

Similarly, brake pressure is remarkably light for an all-disc setup, and fairly weak throttle springs reduce the strain of long distance, high-speed motoring. The engine never seems to be over-worked, and thanks to its aerodynamic shape the Elite gives far less impression of speed than most sports cars — gliding along at 120 mph while others feel to be laboring at 80 mph.

Fortunately, however, this is above all a car which actually *helps* the driver when he comes upon a corner unaware. Whether on the Firestone nylon sports tires fitted as standard, or on Dunlop R5 racing tires, the tail begins to slide relatively early, but completely smoothly and controllably. The Elite can be taken through the majority of corners "on the throttle" with very little movement of the steering wheel and generally the feeling that "I could have gone much faster than that".

The chief difference between Peter Lumsden's Elite and a standard one is that it has considerably more urge at the top end — at the expense of some low-speed flexibility, which means that in traffic it's necessary to shift gears much more often. It is also much more noisy, but this is largely because of the vestigial muffler. A Stage III car with more adequate silencing would be a reasonable enough proposition on the road.

—David Phipps

ROAD TEST

Racing LOTUS Elite

Price as tested: $5700 ($5169 basic)

Importer: Lotus Cars of America, Inc.
4110 Lankershim Blvd.
North Hollywood, California

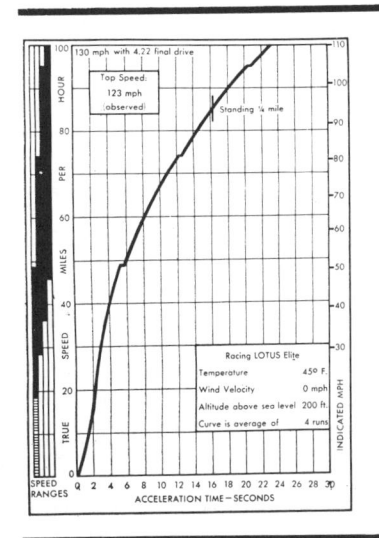

ENGINE: (Coventry Climax FWE)

Displacement	74.25 cu in, 1216 cc
Dimensions	Four cyl, 3.00 x 2.62 in
Compression Ratio	10.5 to one
Power (SAE)	102 bhp @ 7000 rpm
Torque	82 lb-ft @ 5000 rpm
Usable rpm Range	2800-7300 rpm
Piston Speed ÷ √s/b @ rated power	3270 ft/min
Fuel Recommended	Super-premium
Mileage (Le Mans)	20 mpg
Range	335 miles

CHASSIS:

Wheelbase	88 in
Tread, F, R	47 in
Length	150 in
Suspension: F, ind., coil, wishbones incorporating anti-roll bar; R, ind., coil, strut.	
Turns to Full Lock	1⅜
Tire Size	5.00 x 15
Swept Braking Area—disc	302 sq in
Curb Weight (full tank)	1375 lbs
Percentage on Driving Wheels-laden	56%
Test Weight	1705 lbs

DRIVE TRAIN: (BMC B-type close-ratio gearbox)

Gear	Synchro?	Ratio	Step	Overall	Mph per 1000 rpm
Rev	No	3.67	—	16.69	4.5
1st	No	2.45	51%	11.15	6.7
2nd	Yes	1.62	29%	7.37	10.1
3rd	Yes	1.26	26%	5.73	13.0
4th	Yes	1.00		4.55	16.4

Final Drive Ratios: 4.55 to one standard.
3.70, 4.22 and 4.875 to one optional.

The Motor Road Test No. 17/60

Make: Lotus **Type:** Elite
Makers: Lotus Cars Ltd., Delamare Road, Cheshunt, Herts.

Test Data

World copyright reserved; no unauthorized reproduction in whole or in part.

CONDITIONS: Weather: Mild and fine, with light wind. (Temperature 55°-70°F., Barometer 29.9 in. Hg.) Surface: Dry tarred macadam and concrete. Fuel: Super-premium grade pump petrol (approx. 101 Research Method Octane Rating).

INSTRUMENTS
Speedometer at 30 m.p.h.	6% fast
Speedometer at 60 m.p.h.	6% fast
Speedometer at 90 m.p.h.	6% fast
Distance recorder	2% fast

WEIGHT
Kerb weight, (unladen, but with oil, coolant, and fuel for approx. 50 miles) .. 13¼ cwt.
Front/rear distribution of kerb weight .. 46/54
Weight laden as tested .. 16¾ cwt.

MAXIMUM SPEEDS
Flying Mile.
Mean of four opposite runs .. 111.8 m.p.h.
Best one-way time equals .. 113.2 m.p.h.
"Maximile" speed. (Timed quarter mile after one mile accelerating from rest.)
Mean of four opposite runs .. 102.3 m.p.h.
Best one-way time equals .. 104.7 m.p.h.
Speed in gears (at 6,500 r.p.m.)
Max. speed in 3rd gear .. 83 m.p.h.
Max. speed in 2nd gear .. 50 m.p.h.
Max. speed in 1st gear .. 30 m.p.h

FUEL CONSUMPTION
51.0 m.p.g. at constant 30 m.p.h. on level.
54.0 m.p.g. at constant 40 m.p.h. on level.
52.5 m.p.g. at constant 50 m.p.h. on level.
48.5 m.p.g. at constant 60 m.p.h. on level.
43.0 m.p.g. at constant 70 m.p.h. on level.
37.0 m.p.g. at constant 80 m.p.h. on level.
33.0 m.p.g. at constant 90 m.p.h. on level.
29.5 m.p.g. at constant 100 m.p.h. on level.
Overall Fuel Consumption for 2,025 miles, 59.4 gallons, equals 34.1 m.p.g. (8.3 litres/100 km.)
Touring Fuel Consumption (m.p.g. at steady speed midway between 30 m.p.h. and maximum, less 5% allowance for acceleration). 40.5 m.p.g.
Fuel tank capacity (maker's figure) 6½ gallons.

STEERING
Turning circle between kerbs:
Left .. 35 ft.
Right .. 33¾ ft.
Turns of steering wheel from lock to lock 2½

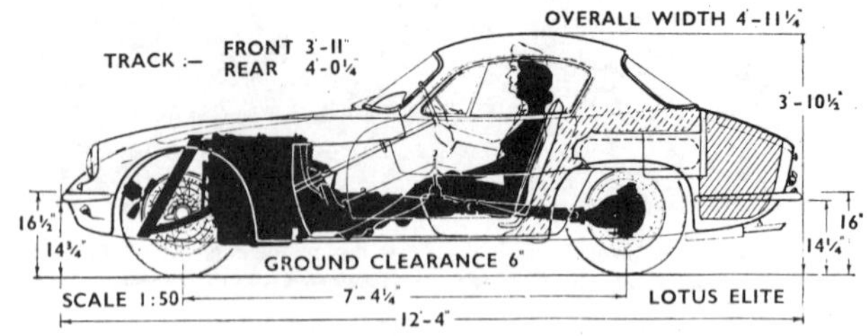

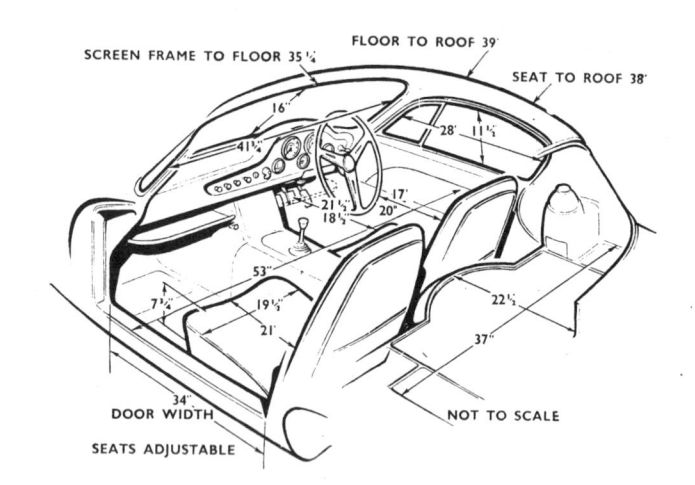

BRAKES from 30 m.p.h.
0.99 g retardation (equivalent to 30½ ft. stopping distance) with 130 lb. pedal pressure.
0.90 g retardation (equivalent to 33⅓ ft. stopping distance) with 100 lb. pedal pressure.
0.65 g retardation (equivalent to 46 ft. stopping distance) with 75 lb. pedal pressure.
0.47 g retardation (equivalent to 64 ft. stopping distance) with 50 lb. pedal pressure.
0.27 g retardation (equivalent to 111 ft. stopping distance) with 25 lb. pedal pressure.

ACCELERATION TIMES from Standstill
0-30 m.p.h.	3.8 sec.
0-40 m.p.h.	5.8 sec.
0-50 m.p.h.	8.0 sec.
0-60 m.p.h.	11.4 sec.
0-70 m.p.h.	15.5 sec.
0-80 m.p.h.	20.4 sec.
0-90 m.p.h.	28.3 sec.
0-100 m.p.h.	41.1 sec.
Standing quarter mile	18.4 sec.

ACCELERATION TIMES on Upper Ratios
	Top gear	3rd gear
10-30 m.p.h.	12.3 sec.	8.3 sec.
20-40 m.p.h.	10.3 sec.	7.4 sec.
30-50 m.p.h.	10.1 sec.	6.7 sec.
40-60 m.p.h.	11.3 sec.	6.5 sec.
50-70 m.p.h.	10.8 sec.	7.8 sec.
60-80 m.p.h.	10.7 sec.	9.0 sec.
70-90 m.p.h.	13.1 sec.	—
80-100 m.p.h.	20.1 sec.	—

HILL CLIMBING at sustained steady speeds
Max. gradient on top gear .. 1 in 9.7 (Tapley 230 lb./ton)
Max. gradient on 3rd gear .. 1 in 6.2 (Tapley 355 lb./ton)
Max. gradient on 2nd gear .. 1 in 3.9 (Tapley 560 lb./ton)

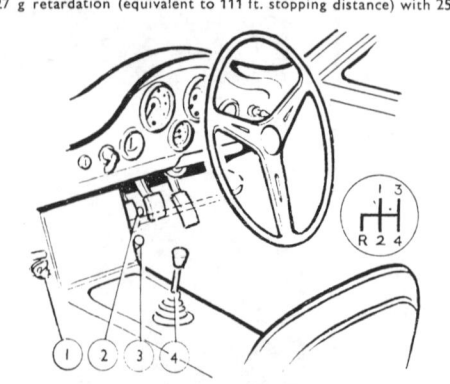

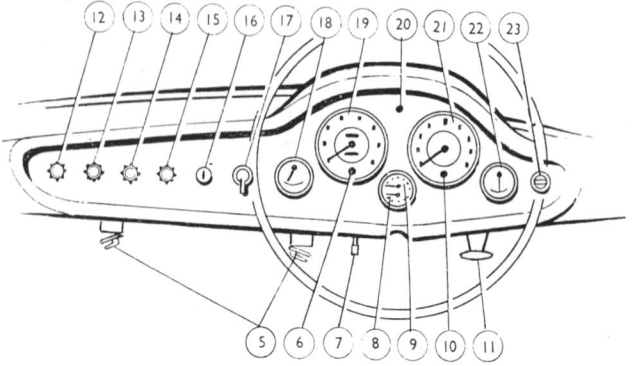

---

1. Heater temperature control. 2. Windscreen washer button. 3. Headlamp dipswitch. 4. Gear lever. 5. Heater air outlet shutters. 6. High beam indicator lamp. 7. Trip re-setting knob. 8. Water thermometer. 9. Oil pressure gauge. 10. Dynamo charge warning lamp. 11. Handbrake. 12. Choke control. 13. Heater fan switch. 14. Windscreen wipers switch. 15. Lights, panel and interior lights switch. 16. Ignition and starter switch. 17. Direction indicator switch. 18. Fuel contents gauge. 19. Speedometer. 20. Direction indicator warning light. 21. Rev. counter. 22. Ammeter. 23. Horn switch and headlamp flasher.

—The Lotus Elite

A Very Small Car Offering Speed, Comfort and Controllability

WHEREAS a great many manufacturers of touring cars have also built racing models, the Lotus organization is altogether unusual in having concentrated upon racing and other forms of competitive motoring for many years before ever attempting to build cars for everyday road use. Even this first " utilitarian " Lotus model, which made its début at Earls Court as a prototype in October, 1957, has been tested-out with tuned eng:nes in a wide variety of races before being put into production as a fully equipped and silenced car for daily business or pleasure motoring. This racing background has produced a car which is right out of the ordinary and, although expensive, a most attractive vehicle.

Throughout the history of the Lotus organization, Colin Chapman has concentrated on producing cars of low weight and wind resistance which had high standards of controllability, so that quite a moderate amount of engine power would provide very high speeds around a racing circuit or from point to point. Without pretending to be very quiet or very weatherproof, many Lotus two-seaters which were built for sports car racing during the past five years proved surprisingly comfortable, and the Elite has not needed to differ appreciably from them in dimensions, suspension layout or power unit type. One big innovation on the coupé model is that, instead of using a multi-tube chassis clothed in metal body panels, it has a chassisless " hull " of polyester resin plastics and glass fibre which combines strength with lightness, smooth contours and quiet, weatherproof comfort.

Fundamental to appreciation of the Lotus Elite is the concept that, although many people require their cars to be fast, comfortable and safe, bigness is no virtue and can, in fact, be a serious disadvantage. The Elite is no larger than its designers have thought necessary for the transportation of two people and their luggage at speeds up to 110 m.p.h. or so, with uncramped elbow room and with racing car standards of road holding. It is a technical tour-de-force to have produced a comfortable, fully equipped car which does this very well indeed for a kerb weight of 13¼ cwt., using a single-carburetter 1,216 c.c. Coventry Climax engine which is working very well within its known limits.

Economy of fuel cannot be regarded as of prime importance for a car costing over £1,900, but the features which have given the Elite its performance have also produced such outstanding petrol consumption figures as 52½ m.p.g. at a steady 50 m.p.h. and 29½ m.p.g. at a steady 100 m.p.h., the light and compact 6½-gallon fuel tank giving a reasonably wide cruising range at 34.1 m.p.g. overall consumption.

Easier to enter than most other very low-built cars, the Elite is certainly a very comfortable two-seater. Individual seats upholstered in leathercloth of high quality have high backrests shaped to provide a reasonable amount of lateral support without making them hard to enter, and sponge rubber cushioning is supported on elastic webbing cross straps. There is plenty of room for long legs and for a straight-arm driving position, quite sufficient headroom, and as much elbow room as could be wished above the central transmission tunnel and in the hollowed-out doors.

Among the few points at which metal is used invisibly to reinforce the body structure is around the windscreen, so that the roof pillars are not so thick as to arouse serious criticism, but the steeply raked windscreen is rather shallow and, whilst a 6-foot-tall driver was exactly satisfied with the seat height, those of greater stature found that upward and forward vision was limited, whereas short drivers would prefer a raised seat to improve vision over the steering wheel rim and the hooded instrument panel.

Early examples of this model which have been seen in races were rather austerely furnished, but the interior trim of current production Elites (the hull of our test car was one which had been moulded by the Plastics Division of Bristol Aircraft) was of very pleasing quality. Inside and out, a discreet touch from the hands of professional stylists has combined neat appearance with honest functional merit. Racing ancestry is recalled by the lightweight wood-rim steering wheel and by the full set of clear, circular-dial instruments, but the touring motorist finds deep and well-fitting pile carpets with rubber inserts, an interior lamp with door-operated switches, variable-brightness instrument lighting, twin windscreen wipers, capacious door pockets, parcel shelves and turn indicators with a time switch to cancel them. Optional extras which were built into our test model comprised a heater and screen demister taking fresh air from an intake on top of the scuttle, and a toe-operated pair of windscreen washing sprays.

Docile Power

Developing 75 b.h.p. in this single-carburetter form as compared with 98-100 b.h.p. when tuned for competitions (with higher compression ratio, twin carburetters, different camshaft and other modifications) the Coventry Climax engine is a potent performer yet very docile. Prompt in starting from cold, it is perfectly happy down to 20 m.p.h. or less in top gear (recording acceleration times from 10 m.p.h. in this ratio seemed no more than mildly cruel) yet gave a timed maximum speed of 111.8 m.p.h. (as the mean of runs in opposite directions) after a rather limited amount of running-in. As the top gear acceleration times recorded on the data page make clear, this engine pulls well at speeds as low as 1,000 r.p.m., but its best torque is not felt until 3,500 r.p.m. is approached, the peak of the power curve is at 6,100 r.p.m., and the red sector on the rev. counter dial starts at a speed of 6,500 r.p.m. which was only just reached in top gear during our timed runs on level road. This is not a silent engine mechanically when the bonnet is open, but a lot of felt and foamed plastic keeps the car interior reasonably quiet, and twin silencers subdue the exhaust note with a completeness which permits unhesitating use of full power on the road at almost any time and place.

Our test car was equipped with a

In Brief

Price with 4.22 axle £1,387 plus purchase tax £579 0s. 10d. equals £1,966 0s. 10d.
Price with 4.55 axle ratio (including purchase tax) £1,949 0s. 10d.
Capacity .. .. .. 1,216 c.c.
Unladen kerb weight .. 13¼ cwt.
Acceleration:
  20-40 m.p.h. in top gear.. 10.3 sec.
  0-50 m.p.h. through gears 8.0 sec.
Maximum direct top gear gradient:
  1 in 9.7
Maximum speed .. 111.8 m.p.h.
"Maximile" speed .. 102.3 m.p.h.
Touring fuel consumption .. 40.5 m.p.g.
Gearing: 16.85 m.p.h. in top gear at 1,000 r.p.m.: 38.6 m.p.h. at 1,000 ft./min. piston speed.

The low, sleek Elite has no cooling apertures to mar its frontal streamlining, even the radiator air intake being used to house the number plate. A hinged panel gives access to the engine bay in which almost every item is easy to reach.

The Lotus Elite

4.22/1 rear axle ratio which is listed as one of several extra-cost alternatives to 4.55/1 gearing. Whilst the latter may be suitable for rallies or for races on slow circuits, the easy and economical speed provided by higher gearing on our test car was something which we would not have wished to sacrifice in order to gain even livelier acceleration in top gear. The four-speed gearbox with central remote control was by no means the car's best feature, third being a good ratio with its maximum of just over 80 m.p.h. but second (in which maximum r.p.m. represented a bare 50 m.p.h.) low enough to be used as a starting-from-rest gear on many occasions: extra running-in would no doubt loosen up a gear-change which was rather stiff, first gear being distinctly awkward to find with the car at rest.

Even with its stiff gearbox and an engine which had enjoyed little running-in, the Lotus Elite can produce very rapid acceleration indeed, from a stand-still or for overtaking from any speed in the range. On a car as light as this our usual test load of two men and their test instruments is a substantial burden, but such figures as 0 to 60 m.p.h. in 11.4 sec., 0 to 100 m.p.h. in 41.1 sec., and 60 m.p.h. to 80 m.p.h. in 10.7 sec., using top gear or 9.0 sec. using third gear, are outstandingly good for so compact and comfortable a closed car.

Silence cannot be claimed for the body interior, there being various resonances from power unit, transmission and road surface which still escape the silencing and sound insulating measures, but the noise which is heard is neither persistent nor ever unduly loud. With the Chapman strut-type independent rear-wheel suspension there are no splined driving shafts to cause transmission snatch, but a vague thud was heard from the hull-mounted final drive unit as the clutch was engaged after changes of gear. Subconsciously, a driver slightly adjusts the throttle opening one way or the other and avoids cruising at the more audible combinations of speed and load, a critical member of our staff who drove and rode more than 800 miles in the Lotus Elite within a 48-hour period being less tired than he would have expected after similarly intensive motoring in almost any other car. The body is well ventilated, with a fresh air heater and screen demister, hinged ventilator flaps of safety glass which really do their intended job, and framed side windows of transparent plastic which lift out completely when a catch is released and can be stowed safely in pockets behind the seats. Thanks to quite exceptionally smooth air flow around the bodywork, removal of the side windows does not result in great noise, draughtiness or appreciable entry of rain during fast driving.

Smooth Riding

Moderately firm springing has been combined with very low unsprung weights to make this car ride exceptionally well over almost any surface. Farm tracks do not worry it (there is reasonable ground clearance despite very low overall height) and the irregularities of fast main roads are ironed out excellently, the shock-free and completely flat ride being very pleasing to passengers. Occasional "bottoming" of the rear springs did, however, raise doubts as to whether the range of spring travel would prove entirely adequate for fast Continental

Two comfortable seats are divided by a high transmission tunnel from which projects the central gear lever. Instruments and fittings are to a keen driver's liking. The seats can be tipped forward to give access to the detachable moulding concealing the spare wheel, above which is a useful shelf between the suspension mounting "domes."

The neat lines of the Elite are evident from any viewpoint; broad but shallow bumpers surround the tail. Although spare wheel and petrol tank protrude into the boot there is reasonably good luggage room.

motoring with the sensibly shaped rear locker really full of luggage.

Extreme sensitivity of the steering and outstandingly good wheel adhesion are the key to the Elite's handling qualities. Collecting it from the factory with incorrectly set tyre pressures, it was brought home to us how sensitive it is to having the correct amount of air in each Firestone 4.80-15 Nylon Sport tyre, and some experiment led to us going somewhat beyond the recommended 20% front/rear pressure difference to use 19 lb. and 25 lb. pressure at front and rear respectively for most motoring. Too soft rear tyres can make the car hesitant directionally when it is being put into a corner or straightened out, although even then it is stable on the straight or when actually cornering. Given the right balance of tyre pressures, it becomes a car which, when tested on the circular steering pad, turns at the same radius for any given steering wheel deflection regardless of car speed, showing neither under- nor over-steer. On the road, rack and pinion steering which feels utterly positive (in spite of a rubber-cushioned universal joint in the steering column which prevents excessive kick-back on rough going) lets the car be steered precisely with incredibly little movement or effort, and even on slippery wet roads remarkably high cornering speeds are needed to provoke a gradual loss of adhesion.

Very smooth control of the car's speed is provided by Girling disc brakes, those at the rear being mounted inboard to save unsprung weight. No brake servo has been thought necessary on such a light car, so these brakes ask for high enough pedal pressures before the wheels will lock to ensure virtual immunity from skidding, but the driver soon becomes supremely confident of their performance on either wet or dry roads, from high speeds or around town. The convenient pull-out handbrake is reasonably powerful, although it would not be safe to leave the car parked on a gradient steeper than 1 in 5.

When a car as unusual as the Lotus Elite comes for test, there is far more which could be said than can be squeezed into four pages. Speed, controllability in all conditions and comfort in all its aspects make this compact two-seat coupé an immensely desirable property for anyone who wants to enjoy covering big daily mileages: expensive in relation to its size and weight, although realistically priced in relation to what it will do and how it does it, the Elite is a perfectly docile runabout for shopping errands and its buyer has no need to budget for a hack "second car" as well as this mettlesome thoroughbred.

The World Copyright of this article and illustration is strictly reserved.
© Temple Press Limited, 1960

Specification

Engine
Cylinders 4
Bore .. .. 76.2 mm.
Stroke .. .. 66.6 mm.
Cubic capacity .. .. 1,216 c.c.
Piston area .. .. .. 28.2 sq. in.
Valves   Single chain-driven overhead camshaft
Compression ratio 10/1
(Optional 11/1 in stage 3 tune)
Carburetter .. One S.U. type H4 horizontal
Fuel pump .. .. .. AC mechanical
Ignition timing control .. Centrifugal
Oil filter .. .. Full-flow
Max. power (gross) .. 80 b.h.p. (75 b.h.p. net)
at .. .. 6,100 r.p.m.
Piston speed at max. b.h.p. .. 2,660 ft./min.

Transmission (as tested)
Clutch .. Borg & Beck 8-in. s.d.p.
Top gear (s/m) 4.22
3rd gear (s/m) 5.57
2nd gear (s/m) 9.28
1st gear 15.49
Reverse 15.49
(Standard final drive ratio is 4.55, with options of 3.7, 4.22 as on test model, or 4.875.)
Propeller shaft .. Hardy Spicer open shaft to sprung final drive unit.
Final drive .. .. .. Hypoid bevel
Top gear m.p.h. at 1,000 r.p.m. .. 16.85
Top gear m.p.h. at 1,000 ft./min. piston speed 38.6

Chassis
Brakes   Girling hydraulic disc type, inboard mounted at rear.
Brake/disc diameters .. .. 9½ in.
Friction areas .. 26.88 sq. in. of lining working on approx. 320 sq. in. rubbed area of discs.
Suspension:
Front: i.f.s. by transverse wishbones, anti-roll torsion bar, and coil springs mounted on Armstrong telescopic dampers.
Rear: Chapman strut-type i.r.s. (Coil springs on Armstrong telescopic damper struts, unsplined double-jointed driving shafts and trailing radius arms.)
Steering gear   Alford & Alder rack and pinion.
Tyres: Firestone nylon high-performance 4.80-15 tubed 4-ply.

Coachwork and Equipment

Starting handle .. .. .. None
Battery mounting .. .. .. In boot
Jack .. .. .. Lazy-tongs type
Jacking points .. 2 points, one under each side of body
Standard Tool kit: Jack and handle, 3 spanners, adjustable spanner, sparking plug spanner, pliers, screwdriver, copper wheel nut hammer.
Exterior lights: 2 headlamps (Lucas 7-in. "Le Mans"), 2 sidelamps, 2 stop/tail lamps. 2 number plate lamps.
Number of electrical fuses .. 1 plus 1 circuit breaker in lighting switch.
Direction indicators .. Amber flashers with self-cancelling time switch
Windscreen wipers .. Lucas two-speed self parking, with twin blades.
Windscreen washers .. .. Optional extra (Tudor toe-operated)
Sun visors .. .. None
Instruments: Speedometer with total and decimal trip distance recorders, rev. counter, oil pressure gauge, coolant thermometer, ammeter, fuel contents gauge.
Warning lights   Dynamo charge, headlamp main beam, turn indicators

Sump .. .. 8 pints, S.A.E. 20 winter, S.A.E. 30 summer
Gearbox .. .. 4½ pints, S.A.E. 40 winter, S.A.E. 50 summer
Rear axle .. 1¾ pints, S.A.E. 90 hypoid gear oil
Steering gear lubricant .. .. grease
Cooling system capacity .. 12 pints (1 drain plug)
Chassis lubrication   By grease gun every 1,500 miles to 14 points
Ignition timing .. 2°-3° before t.d.c. static
Contact-breaker gap .. 0.014-0.015 in.
Sparking plug type .. .. Champion N3
Sparking plug gap .. .. .. 0.018 in.

Locks:
With ignition key ..Ignition/starter switch, driver's door, luggage locker
With other keys.. .. .. None
Glove lockers .. .. .. None
Map pockets .. Two inside doors
Parcel shelves   One behind seats, shelves below facia integral with ducts for optional heater
Ashtrays .. .. None
Cigar lighters .. .. .. None
Interior lights .. .. One in roof with courtesy switches
Interior heater .. Optional extra Smiths
Car radio .. .. .. None offered
Extras available: Interior heater, windscreen washers, special colours, seat belts, Lotus stage 2 or stage 3 engine tuning modifications, close-ratio gears, and other racing equipment.
Upholstery material .. .. Leathercloth
Floor covering .. .. .. Carpet
Exterior colours standardized 3
(Any other colour at £35 extra)
Alternative body styles .. .. None

Maintenance

Valve timing   Inlet opens 12° before t.d.c and closes 56° after b.d.c; Exhaust opens 56° before b.d.c. and closes 12° after t.d.c.
Tappet clearances (Cold).. .. Inlet 0.006 in Exhaust 0.008 in
Front wheel toe-in .. .. 1/16 in. to 1/8 in.
Camber angle .. .. .. 1¼° to 1½°
Castor angle .. .. 7°
Steering swivel pin inclination .. 9°
Tyre pressures   Front 19 lb. Rear 23 lb. (see text Raise pressures by 7 lb. for very fast driving.
Brake fluid .. .. .. Girling crimson
Battery type and capacity   12 volt, 57 amp. hr
Miscellaneous .. Car should be jacked up ONLY by jacking points provided.

EXCELLENT FINISH is a feature of the low-drag wind-cheating grand tourer. Ample luggage room and remarkable economy are among the attractions of this delightful vehicle.

JOHN BOLSTER TESTS
THE LOTUS ELITE

This body is supported on four independently sprung wheels. In front, there are wishbones and helical springs with telescopic dampers, reinforced by a torsional anti-roll bar. The steering is by rack and pinion. Behind, the Chapman strut-type suspension employs the articulated half shafts to locate the hubs laterally. The struts incorporate helical springs and telescopic dampers, and there are trailing arms for fore and aft location. Girling disc brakes are used all round, the rear ones being inboard mounted at either side of the differential.

Although the car is very low, entry and exit through the large doors is easy. The front sections of the windows swivel for ventilation, the main positions having no sliding mechanism but being instantly removable. The driving position is

EVERY reader of AUTOSPORT must be familiar with the performance of the Lotus Elite in competitions. At any race meeting we now only show the faintest surprise when the little 1,216 c.c. coupés overwhelm the 3½-litre brigade, and when they out-corner other G.T. cars we think it hardly worthy of comment. As a competition car, the Elite has most certainly arrived.

Yet, the virtues which show up so well on the circuits are perhaps even more valuable on the road, and it is as a fast touring car that I have recently been using an Elite. The engine was in standard, single-carburetter form, but could at any time be brought up to racing specification if the owner wished to engage seriously in competition work. Similarly, the gearbox had touring ratios instead of the more closely spaced speeds that are desirable when racing is afoot.

The ultimate performance of "my" car was, therefore, below that of the ones supplied to "the boys", in the interest of flexibility, silence and a reasonable purchase price. It had, however, the full road-holding and braking qualities, and, of course, the low-drag body form was identical. In standard form, the Elite is a small but practical fast touring car for two, with considerable luggage space, remarkable fuel economy, and a delightful appearance. The finish is really excellent, the mechanical organs are accessible, and a really full tool kit is supplied.

The basic structure of the Lotus Elite is a body-cum-chassis of glass-reinforced polyester resin. This material is particularly suitable for such a fuselage, because the thickness may be graduated according to the stresses to be resisted by every part. For a car in limited production, this avoids the almost countless pressings that are welded together to form the modern steel saloon, and the plastic is less inclined to transmit and magnify noise than would be a metal-panelled shell.

A Coventry Climax light alloy engine is mounted together with the gearbox at the front of the car. A Hardy-Spicer propeller shaft transmits the drive to the chassis-mounted hypoid unit, which had a 4.22 to 1 ratio in the case of the test car. The ratio may be specified to choice, but 4.55 to 1 is the usual wear, giving slightly brighter acceleration but perhaps a small drop in maximum speed. In so low a car, a very deep propeller shaft tunnel is unavoidable. However, it has the advantage of locating the driver and passenger during fast cornering. The two bucket seats are also profiled for this purpose, and have folding backs for access to the rear of the car. The spare wheel is carried horizontally inside the body, concealed by a detachable cover, while the capacious boot is reached from outside by a normal lid. There is also plenty of pocket and shelf space.

superb, as one would expect, the steering wheel, with its thin wooden rim, being ideally placed, and the pedals are correctly located for "heel and toe". The windscreen pillars are not of the thinnest and the roof is slightly beetle browed at the corners of the screen, but in practice no objectionable blind spot is found to exist. The all round visibility is good, and the car has a light interior. The short gear lever projects conveniently from the central tunnel.

Regarded as a touring car, the success of a machine of this calibre must depend largely on the noise level inside. The Elite is unusual, inasmuch as it is rather noisy when driven slowly and manœuvred on the lower gears, but strikingly quiet at high cruising speeds. The engine and gearbox are very well insulated from the occupants, such noise as one hears being "telephoned" down the propeller shaft and communicated to the structure via the final drive unit. Thus, one has the curious sensation of riding in a rear-engined car, though the

DRIVING POSITION is superb, all-round visibility is good, and access through the large doors is easy.

SPECIFICATION AND PERFORMANCE DATA

Car Tested: Lotus Elite 2-seater coupé, price £1,949 including P.T.

Engine: Four cylinders 76.2 mm. x 66.6 mm. (1,216 c.c.). Single overhead camshaft directly operating valves through inverted piston tappets. Light alloy head and block. Compression ratio 10 to 1. 75 b.h.p. at 6,100 r.p.m. Single 1½ ins. SU carburetter. Lucas coil and distributor.

Transmission: Single 8 ins. dry plate clutch. Four-speed gearbox with short central lever and synchromesh on upper three gears, ratios 4.22, 5.57, 9.28, and 15.49 to 1. (Alternative final drive ratios 3.7, 4.55 and 4.875 to 1. Chassis-mounted hypoid final drive unit.

Chassis: Integral chassis/body construction of glass-reinforced polyester resin. Independent four-wheel suspension by helical springs and telescopic dampers; front suspension by wishbones, anti-roll bar and rack and pinion steering; strut-type rear suspension employing articulated half-shafts for lateral location with trailing arms. Girling hydraulically operated 9½ ins. disc brakes, inboard at rear. Knock-on wire wheels fitted 4.80-15 ins. Firestone tyres.

Equipment: 12-volt lighting and starting. Speedometer. Rev. counter, oil pressure and water temperature gauges. Ammeter. Two-speed self-parking windscreen wipers and washers. Flashing indicators. Extra: heating and demisting.

Dimensions: Wheelbase, 7 ft. 4 ins. Track, 3 ft. 11 ins. Overall length, 12 ft. 6 ins. Height to roof, 3 ft. 10 ins. Width, 4 ft. 10 ins. Ground clearance, 6½ ins.

Performance: Maximum speed, 108.19 m.p.h. Speeds in gears: 3rd, 87 m.p.h.; 2nd, 54 m.p.h.; 1st, 34 m.p.h. Standing quarter-mile, 17.4 secs. Acceleration, 0-30 m.p.h., 2.8 secs; 0-50 m.p.h., 6.8 secs.; 0-60 m.p.h., 9.8 secs.; 0-80 m.p.h., 17 secs.

Fuel Consumption: Driven hard, 30 m.p.g. Driven moderately, 40 m.p.g.

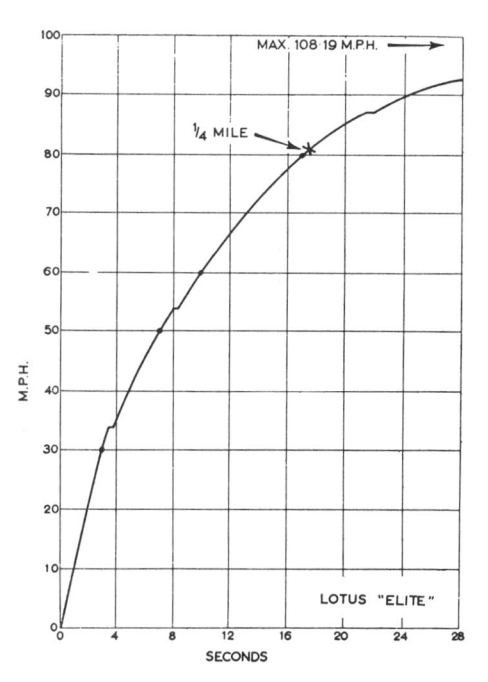

ACCELERATION GRAPH

power unit is, in fact, at the front. The absence of exhaust noise must be praised.

As the performance figures show, the Elite is a lively car even in touring tune. All the power may be used *all* the time, for even full throttle getaways on wet and slippery roads fail to produce any wheelspin, thanks to the excellent independent rear suspension. There is a comparatively large gap between the second and third speeds, but the wide revolution range of the engine tends to mask this. At well over 80 m.p.h. in third gear, the power unit is still perfectly happy.

On top, the car seems to smooth right out at over 90 m.p.h., and will cruise in delightful ease at a genuine 100 m.p.h. We do not, of course, publish average speed claims in AUTOSPORT, but it is perhaps permissible to remark that I was able to average more than 100 m.p.h. for the full length of M1. At the timed maximum speed of just over 108 m.p.h., the piston speed is only 2,650 ft./min. so there is not the slightest harm in allowing the car to run up to 120 m.p.h. on down grades or when the wind is astern.

The road-holding is quite outstandingly good and the cornering power is very high indeed. The steering is very light and direct, so the whole process of negotiating curves rapidly is a matter of gentle wrist movement. Changes of surface and camber may be almost ignored, and the sensation of absolute safety, which is felt by the driver and passenger alike, must be experienced to be believed.

This phenomenal road-holding is matched by superb brakes. The Girling discs are most unspectacular, for there is no screaming of tyres or slewing sideways as the car is brought to an emergency stop. Even at three-figure speeds, there is no tendency for the wheels to lock, and the combination of fade-free discs with independent four-wheel suspension results in retardation of a rapidity which no ordinary car could approach.

POWER from the 75 b.h.p. Coventry Climax engine will propel the Elite at up to 108 m.p.h. in standard form, but very much "hotter" units are available for competition work.

The ride of the Elite is firm at low speeds, smoothing right out as the natural cruising speed is reached. It is a level ride, with no suspicion of pitching, and the suspension makes no noise as the wheels react to the bumps. The seats are not sprung, so one feels the small surface irregularities at town speeds, but the big bumps are always swallowed. Certain types of road surface transmit some noise into the car, but this is not excessive.

The handling characteristic is as near neutral as may be, and no unexpected rear-end breakaway is ever experienced. It must be made clear that the whole pattern of handling and controllability is literally in a different world from that of the typical popular sports car. Yet, the Elite presents no new handling problems. The most uninspired driver will put up a better show in this car, and be safer while he is doing it, than in his own dreary conveyance. The better driver will at once sense that the Lotus is responding to his very thoughts as no car ever has before. It is true that the Elite is in some respects like a racing car to drive, but it is much more forgiving than the modern single-seater and gives the driver full warning when the limit is being approached.

This *rapport* between the car and the driver is difficult to put into words. Anybody who has gone through a series of curves on a really good racing motor-cycle may know that sensation when bike and man appear to become one. In Greek mythology there were gentlemen called centaurs who were half man and half horse, and I am sure that I became half Bolster and half Lotus as I flicked the Elite through the corners. There are bigger and faster sports cars which remain strangers to the driver, and in consequence he is unable to put up high average speeds in safety.

The Lotus Elite is not a cheap car, but there is nothing quite like it for fast, safe motoring.

If two seats are enough, and sheer motoring enjoyment is the object, it is difficult to imagine how £1,949 could be better invested.

The LOTUS ELITE

A TECHNICAL APPRAISAL

BY CHARLES H. BULMER

OF COLIN CHAPMAN'S NON-METALLIC G.T. COUPE

THREE years ago when the decision was made to develop a Grand Touring Lotus for quantity production, this type of car was hardly manufactured in Britain except in the heavier, large-capacity classes. The Lotus company was in the fortunate position of having a number of chassis and suspension components highly refined by years of racing and perfectly suited to such a car provided the weight could be kept low.

The existence of the 1,100 c.c. Coventry Climax, with its astonishing power-to-weight ratio, was obviously a major factor in establishing both the practicability of the project and the scale of the car. Without alteration of other components, it was originally intended to use a slightly cheaper version of the engine with cast iron block and crankcase and a capacity of nearly 1,300 c.c., but this has never materialized and all Elite engines still retain the light alloy block. They are, however, bored out to 1,216 c.c. and, fitted in standard form with a single 1½ in. S.U. carburetter, give 75 b.h.p. (gross) at 6,100 r.p.m. Much more highly tuned versions are available for competition purposes, but this engine is now so well known and documented that it will not be described here in detail. It is attached to the M.G. Magnette version of the equally familiar B.M.C. "B" type gearbox either with standard ratios or the special close ratios that are available as an optional extra.

The original planning committee comprised Colin Chapman as chairman and structural engineer, John Frayling and Peter Kirwan-Taylor on the styling side (the former sculpted the models) and Frank Costin for aerodynamic advice. The body shape that finally emerged from their sometimes conflicting requirements was never wind tunnel tested at the design stage, although Costin made some low-speed airflow experiments with models, and the result is a demonstration of what can be done by a combination of scientifically informed guesswork and a careful regard to some of the less obvious parts like the undertray and the cooling system. Full-scale tunnel tests made since suggest a drag coefficient in the region of 0.29 and although this figure should perhaps be treated with reserve because of the difficulty of simulating ground effects in tunnels, the maximum speed obtained in our Road Test (May 11, 1960) suggests that the true value cannot be greatly in excess of 0.3 compared with the figure of 0.5 which is more representative of most modern cars. At a steady 100 m.p.h. the fuel consumption was as low as 29.5 m.p.g.

Rigid Plastic Structures

Whilst it was clear that these cars could not be built at a reasonable price with hand-made sheet metal bodies, it was equally clear that, for the sales envisaged, the cost of dies for pressings could not be contemplated, so that the choice of a plastic body was almost automatic. This could have clothed a separate chassis frame of some sort, but it was felt that the Lotus tradition of minimum weight and maximum number of functions for any given part would be violated by a design which allowed all this structurally well-positioned plastic material to escape unstressed, and the decision was made to build an integral body/chassis unit mostly in glass-reinforced polyester resin.

This material has properties very different from the usual metals. With approximately one-third of the tensile strength of steel, but only one-fifth of the weight, there is no difficulty in producing structures of high strength to weight ratio. But with reasonable design, strength is seldom the limiting criterion of a chassis and it often follows automatically on the attainment of adequate stiffness in bending and torsion. In this respect the material is at a disadvantage since the modulus of elasticity, which represents the amount of elastic stretch or distortion which corresponds to a given stress, is only about 6% of that of steel, and the stiffness to weight ratio of a structure like this may be reduced by a factor of nearly 4, other things being equal.

Fortunately, other things are not entirely equal; one of the hidden assets of this form of construction is the ease of varying the thickness and shape of the material to accord with the local loads, and with clever design the weak points, which may account for much of the total flexibility, can be greatly reinforced. Basically it can be seen that at the front of the car the scuttle structure, the wheel arches and front wings and the nose cowling all combine to form a box structure of immense rigidity. Similarly the wings and wheel arches, the boot, the rear bulkhead and the spare wheel tray form another inherently rigid closed structure at the back.

Stiffening the Centre

The real problem is to join these ends with an equally stiff centre-section despite the enormous cut-outs for the doors and windscreen. At floor level the various mouldings combine to form a very deep central rectangular tube (containing the propeller shaft) and torque boxes of roughly triangular cross-section running under each of the door sills. The roof panel, of course, is the other vital bridging structure with considerable inherent rigidity but joined to the front part of the car only by two windscreen pillars which must be as slender as possible for the sake of visibility. Tests on a prototype showed that much of the car's overall stiffness was lost here, torsional loads causing lateral bending of these pillars whilst their fore and aft bending permitted vertical flexing of the shell; the pillars are now reinforced by bonding-in the steel hoop structure, shown in the drawing, welded to square vertical tubes which take the door hinges and provide jacking points at their lowest extremities.

This kind of development has raised the torsional rigidity to approximately 3,000 lb. ft. per degree of chassis twist (measured between the fore and aft wheel planes), a figure which although not unusually high by current standards of unit steel construction, is more than adequate in relation to the light weight of the car and its major components, and on the road it certainly feels outstandingly rigid.

The only other metal component of any size is the steel subframe which is bonded to the plastic front structure and carries the wishbones, the anti-roll bar, the steering rack and pinion, and the engine front bearer bolts. Including this subframe, the windscreen hoop, the doors, bonnet and boot lid, but without glass or interior trim, the whole of this body/chassis unit weighs about 300 lb.

Experience has shown that it is quite unsatisfactory to bolt rigid components directly to the plastic structure. Compression of the resin allows local fretting, and crumbling of the glass reinforcement brings increasing looseness. By interposing rubber in compression between the two parts this trouble can be avoided, and the engine, differential casing and rear suspension are all mounted by means of rubber bushes or rubber sandwiches. Where metal parts have to be bonded to the plastic directly, no reliance is placed on adhesion between the two, but by liberal drilling of the metal or other means a mechanical key is formed.

Initially, a batch of 25 experimental Elites was built and sold to selected competition drivers who could be relied upon to uncover any inherent weaknesses. It was soon found that the

Left: An inverted view of the undertray section, showing the hole for the sump and the moulded channels for the exhaust system which bifurcate at the back just in front of the differential mounting.

Below: The floor moulding being bonded to the undertray. As the former is narrower, it can be seen that open "trough" sections are left at each edge under the door openings. These are closed by the third moulding to give boxes of great strength.

The LOTUS ELITE

intense heat generated by the inboard disc brakes under racing conditions caused local softening of the resin and eventual failure of the differential mounting. The trouble was later eliminated by the use of heat shields and by strengthening of this section which now has a material thickness of up to 0.7 in. compared with the average of about one-eighth of an inch, diminishing to less than one-tenth of an inch in lightly stressed areas.

Forming the Structure

The various body components are built to the required shape by forming them in female moulds. These are first coated with a parting agent, to prevent the finished product adhering permanently, followed by a thin coat of resin, called a "gel coat," which is allowed to dry and which ensures that a smooth finish is presented, free of flaws, blowholes and glass texture. Onto a further thick, wet coat is laid the glass reinforcement in the form of a thin mat of randomly arranged short glass fibres which is pressed well into the former and thoroughly impregnated with more resin to remove trapped air. Several layers may be built up in this way depending on the thickness required, and finally, to accelerate the curing process, the whole is moderately heated for a few hours before removal from the mould.

The main structure comprises three large separate mouldings and a number of smaller subsidiary ones which are finally bonded together. The foundation section is the undertray, which embodies the front and rear wheel arches, the differential mountings, and some of the rear suspension attachment points, and to which is bonded the fabricated steel sub-frame for the front suspension and steering. The second moulding includes the floor, the deep propeller-shaft tunnel and the spare wheel tray, and a photograph shows the fixture in which it is being bonded to the separate bulkhead and the undertray which closes in the fourth side of the central tunnel. A further illustration shows the remaining part of the main structure comprising the upper parts of the body again inverted for insertion of the inner roof panel. Bonding of these components over their contact areas with local pressure fixtures produces a shell which encloses numerous box sections and which is double-skinned throughout so that only the presentable surfaces are visible, greatly improving the finished appearance of the car. In similar fashion the doors, bonnet and boot lid are moulded in separate inner and outer skins (using the stronger epoxide resin) and assembled to produce hollow sections of great rigidity. It is impossible to watch the assembly of these bodies without being greatly impressed by the structural forethought and detail planning that has gone into the design.

The bodies are finally delivered to Lotus with all glazing in place, doors mounted, fully painted and complete with interior trim and sound damping. At present the Plastics Division of Bristol Aircraft, Ltd., is producing them at the rate of over 100 a month and it may well be that the amount of handwork involved will limit the economic field for glass construction to production rates of not more than two to three times this number; the figure is very dependent on possible improvements in production technique and on the value the designer sets on his greater freedom to abandon the design and scrap the relatively inexpensive manufacturing tools when this is considered technically desirable.

Turning now to the suspension, the strut assembly used at the back, which made its first appearance on the original Formula 2 Lotus, has caused a degree of confusion in the past, some writers referring to it as a swing-axle system, but true swing-axle geometry, as exemplified by the Volkswagen and Renault Dauphine, is characterized by the use of only one universal joint in each half-shaft.

The Lotus rear suspension, like the Macpherson front suspension of current Ford cars, is geometrically related to the large family of double-wishbone systems, the lower wishbone being formed by the fixed-length half-shaft and the triangulated radius arm. If the Armstrong spring/damper unit were of fixed length instead of being telescopic, and if the upper mounting

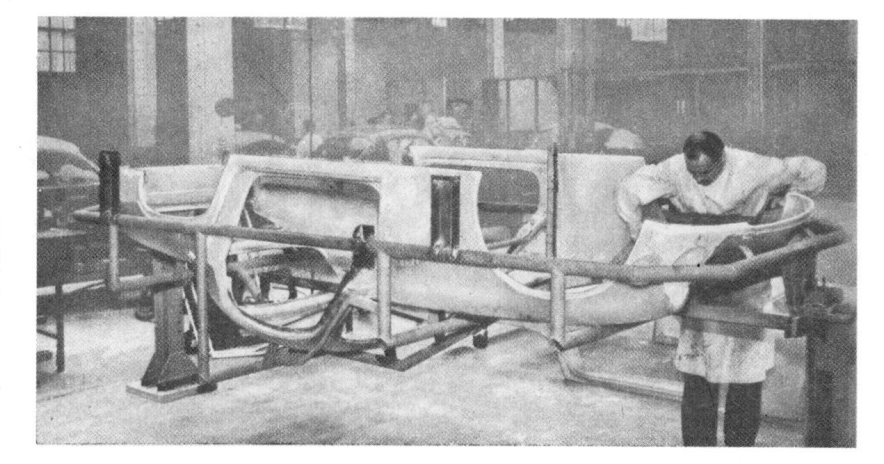

The upper parts of the body are shown inverted in a fixture whilst an inner panel is bonded to the roof. Through the front wheel opening can be seen the square door pillar which is welded to the windscreen reinforcing hoop.

(B in the accompanying diagram), instead of being bolted to the body shell were constrained by a link or wishbone coincident with the line AB, it will be seen that the motion of the wheel would be little affected for small movements, although exact equivalence would demand an upper wishbone of infinite length. Those who are interested in suspension geometry will see from this diagram that in bump and rebound the wheel moves about an instantaneous centre at A, so that the roll centre (RC) is at the intersection with the plane of symmetry of the line joining A to the tyre contact point. This centre is actually about 6 in. above ground level in the normal loaded position and the rear wheels have about 2° negative camber.

From the mechanical point of view, the advantages of the design include the ingenious duplication of functions which has reduced the number of parts to a minimum, and the way that the loads applied to the structure are reduced by wide separation of the attachment points. Disadvantages include the height of the suspension strut, which imposes a limit to the wheel movement which can be provided, and the considerable bending moments applied in certain load configurations which can greatly increase the normal sliding friction of these units, although this has caused no difficulty with the Elite.

In the earlier cars, a cranked longitudinal radius arm gave fore and aft location of the rear wheels and also, by means of a forked end engaging with the lower part of the wheel carrier, kept them pointing in the right direction. Because of the narrowness of this fork it was necessary to use taper roller bearings for precision of control and even then bending of the arm, together with some tendency for the fork to spring, made this rather expensive construction less than satisfactory. It is not widely appreciated what accuracy of control is needed to prevent unwanted steering effects, but as an illustration, 1° of lock on both rear wheels would steer the car sharply enough at 80 m.p.h. to put it into a slide. In the revised design, seen in the sectioned drawing, the single arm has become a triangulated wishbone. Adequate separation of the two bearings on the hub carrier has enabled rubber bushes to be used, and the forward mounting on the chassis has been moved considerably inboard, practically eliminating the previous wheel toe-out on bump and rebound movements.

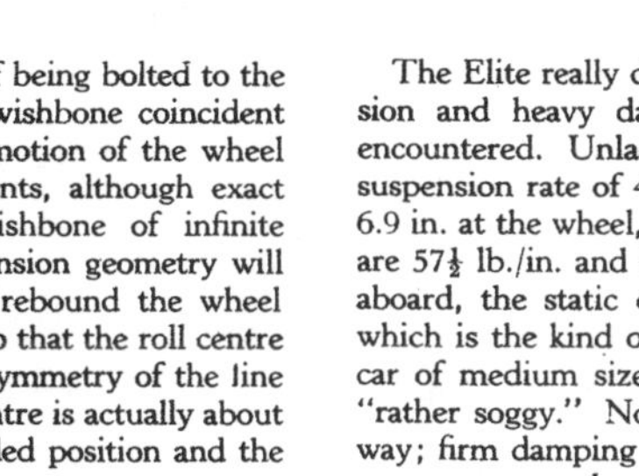

The front suspension is of double wishbone design using an Armstrong coil spring/damper unit and an unusually large and powerful anti-roll bar which forms one member of the upper wishbone and increases the overall front roll stiffness by a very large factor. The wheels are set at zero camber angle, the roll centre is about 4 in. above ground level, and a total wheel travel of 6 in. is provided. Although the 9½ in.-diameter Girling brakes are outboard at the front, the unsprung weight (54 lb. each side) is only 1 lb. greater than at the rear where the hub carrier, bearings and half-shafts compensate for the inboard mounting of the brakes.

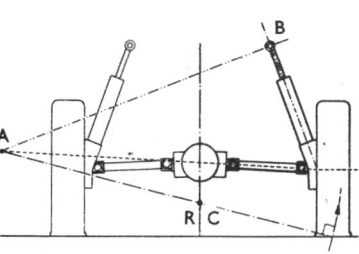

The geometrical construction for determining the position of the roll centre of the rear suspension.

The Elite really does embody the combination of soft suspension and heavy damping that is often extolled but seldom encountered. Unladen, but otherwise ready for use, the front suspension rate of 42 lb./in. gives a nominal static deflection of 6.9 in. at the wheel, whilst at the back the corresponding figures are 57½ lb./in. and 6 in. In touring condition, with two people aboard, the static deflections increase to approximately 8 in., which is the kind of figure one expects in a comfortable family car of medium size—the sort one might describe unkindly as "rather soggy." No one could possibly describe the Lotus this way; firm damping eliminates all trace of wallowing, and many competent people who drove this car on test thought it combined excellence of roadholding and ride more successfully than any car in their previous experience.

The Designer's Headache

The very low centre of gravity and high roll stiffness have almost eliminated the brake dive and cornering roll that might have resulted from the soft springing, but load variation at the rear is still a designer's headache. A total wheel travel of 7 in. is available, but two people and some luggage can easily increase the sprung weight at the rear by one-third and lower the static position 2 in. Progressive rubber stops are now used to absorb some of the load well before the end of the travel, but of course this will also lead to a rising rear roll stiffness.

The steering, by Alford and Alder rack and pinion, is conventional in layout but, in common with several other high performance machines, the orthodox Ackerman geometry has been abandoned. It is now recognized that this traditional layout will give an approximation to correct steering angles only at very low cornering loads when the tyres are running at negligible slip angles and it is hardly likely that Lotus owners will drive in this unenterprising way. The Elite, however, is unusual in having gone a stage further than keeping its wheels parallel on lock and uses negative Ackerman effect so that the outer front wheel on a corner turns through a greater angle than the inner one. Whilst this is undoubtedly correct for fast cornering, it would be interesting to know whether it has any connection with the unusual sensitivity that the steering displays in other conditions.

In this description it has only been possible to touch on a few of the many interesting facets of a car which has been evolved by a team of singular ingenuity and analytical ability with little regard for convention. As regards any future models they may design, it is most encouraging to find that they drive their own product far and fast enough to be more aware of its faults than anyone else.

The World Copyright of this article is strictly reserved.
© *Temple Press Limited, 1960*

ROAD IMPRESSION

THE LOTUS ELITE

WE have just had the good fortune to complete over 1000 miles in a Lotus Elite—a much talked-about sports car of unique design. It was 1000 miles of traffic driving, stop and start motoring, long distance country runs and high speed testing: 1000 miles at an average fuel consumption of 36.5 mpg, accelerating from 0-100 mph in less than 35 seconds and attaining a maximum speed of 110 mph. And what do we think of the Lotus Elite? That question can be answered in one short sentence—'it puts the fun back in motoring.'

GLASS FIBRE BODY

As is already well known, the Elite is unique in that it is almost wholly constructed of glass fibre. There is no chassis frame as such, since the body and chassis are made in one unit from several glass fibre mouldings. There are two sub-frames moulded into the plastic, one of which re-inforces the roof structure and, at the same time, provides a strong point for jacking. The other is built into the front of the car, providing attachment points for engine and front suspension.

The power unit of the car under test was the Stage I version of the 1216 cc Coventry Climax engine, which gives 75 hp at 6100 rpm. Compression ratio of this single carburetter version is 10 to 1 and yet, amazing as it may seem, it retains complete flexibility. Transmission is through a normal four speed BMC "B" type gearbox, fitted with the same gears as the MG Magnette saloon, to a hypoid final drive unit fitted as standard with 4.2 to 1 crown wheel and pinion.

The rear suspension, which has become known as "Chapman strut" type, uses the double articulated half shafts to provide lateral location, with combined coil spring damper units as the suspension medium. Longtitudinal location is looked after by short "A" shaped trailing arms. Front suspension is independent by transverse wishbones, which incorporate an anti-roll bar, and combined coil spring and damper units.

Disc brakes are fitted all round, those at the rear being mounted inboard on the final drive unit. The diameter of the front discs is 9½in. Steering is by rack and pinion gear.

We took over the Elite at the Lotus factory at Cheshunt, with a full tank of fuel and instructions to test it to the utmost, and not to bring it back for 10 days. The first impression gained, whilst driving into London, was one of complete docility. This car could be treated like any other small saloon. It could be driven in top through the heaviest traffic, and only if the speed dropped below 20 mph was it necessary to change down into third. After the first few miles, learning the whereabouts of the controls and getting used to the feel of things, there was ample time to have a look round the interior of the Elite whilst patiently waiting in the traffic hold-ups which abound when travelling in London.

COMFORTABLE SEATS

The high backed seats are luxuriously comfortable whilst the seat cushions, unlike so many modern production cars, provide plenty of support for the thighs. Perhaps the seat is a little on the low side for a driver of 5ft. 8in. or less, so that the rim of the steering wheel obstructs, somewhat, the view of the road. A cushion on the seat soon cured this and obviously an owner suffering from the same problem would raise the seat an inch or so. The angle of the steering wheel itself is perfectly comfortable.

The instrument panel is neatly set into the all plastic facia and the dials themselves are provided with a cowl to prevent reflection in the steeply sloping windscreen. The instruments are plain and functional, as they should be, and are very easy to read. The speedometer reads to 140 mph, whilst the tachometer (rev. counter) reads to 8000 rpm. Other instruments are a combined oil pressure and water temperature gauge, an ammeter and a fuel gauge. The controls and switches, which are all positioned conveniently for left handed use, consist of (reading from left to right) a pull-out choke control, heater, 2-speed wipers, lights, ignition and flashing indicators. To the right of the instruments is a stalk-type two-way switch, which, when lifted, flashes the main beam of the headlights, and when depressed sounds the horn.

Centre of attraction inside is the delightfully short gear lever which protrudes from the top of the deep transmission tunnel. The normal ratchet-type hand brake lever is at the side of the tunnel, and both controls are perfectly situated for any driver.

We particularly liked the removable side windows. Since these cannot be arranged to wind down as in conventional saloon cars, the designers have considered it wiser to have them completely removable rather than fit sliding windows, which are so often fitted to sporting coupes. The quarter lights are not removable, and, with these open, and the windows stored in their pockets behind the seat squabs, there is no draught whatsoever. This indicates that the Elite has an efficient aerodynamic shape.

The inside of the roof does perhaps look a little stark, since it consists of uncovered plastic. However, it is finished in a pleasant matt grey and is obviously much easier to keep clean than is any fancy upholstery.

It is when the towns have been left behind that one realizes how much the Elite is a "wolf in sheep's clothing." Now it becomes

The clean uncluttered lines of the Lotus Elite are praiseworthy, for it manages to look attractive from every angle.

a high speed carriage for two persons, capable of covering the miles in as short a time as any other vehicle on the road, bar none.

When starting from rest, it is normal to use bottom gear (although it will move off quite smoothly using second), accelerating to about 4500 rpm before changing into second gear. Once in this gear, acceleration is very rapid indeed and 6000 rpm—50 mph—comes up in no time at all. Third gear can be held quite comfortably until well over 80 mph is attained before changing into top. At this speed one feels quite happy that this is the car's normal cruising speed. Conversation can be carried on quite normally since the engine, although turning over at about 5500 rpm, is running as smoothly as a dynamo.

The low-geared steering is very sensitive, so that fast main road bends are taken without consciously turning the wheel. It is on the secondary roads that the road holding and steering of the Elite really begins to show up that of other motor cars. Corners which one would normally take at about 30 to 40 mph may be taken at 50 to 60 mph in absolute safety. There is no tendency to deviate from the chosen line by more than an inch or so, and that is due to a slight trace of understeer which is just noticeable when accelerating through these bends.

For those who do not like a car with understeering characteristics, however slight, it should be pointed out that the tyres on our test car were inflated to the recommended pressures of 19 lb in the front and 23 lb in the rear. An increase in pressure at the front

Elite engine compartment with the Climax unit and a single carburetter.

or a decrease at the rear would quite obviously help to promote slight oversteer.

On smooth surfaces road noise transmitted to the car is very slight, and only on rough or bumpy roads is one aware of tyres revolving and suspension working. Even then, this suspension smooths out every undulation—it is heard but not felt. It takes only a few miles of fast driving to accustom yourself to this car and then, provided it is driven with the head and not the feet, it is in our opinion the safest motor car on the roads today.

The Girling disc brakes, which are not servo-assisted, give an immediate feeling of security, especially when they are applied at anything over 80 mph. At this speed, they respond at the slightest touch of the pedal, and, if an emergency stop is required, they pull the car up in a perfectly straight line without wheels locking or any fuss whatever. At slow speeds, more pedal pressure is required than would be expected, but this is a characteristic of non-servo disc brakes.

One reason for the effectiveness of the brakes is that adhesion of all four road wheels is particularly good. During acceleration tests it was found impossible to make the rear wheels spin. As the clutch was let in at about 4000 rpm, the rear of the car can be felt "digging in" to the road surface.

On the road, this particular Elite could be faulted on one point only. The engine appeared to have a slight 'out-of-balance' period between 3800 and 4200 rpm. This could be felt and heard, to some considerable extent when accelerating hard, as piston slap or little end knock. However, when cruising at a steady 4000 rpm in top (about 63 mph) on a trailing throttle, it was hardly noticeable. Again, pinking could be induced by vicious acceleration despite using the very top quality grade of fuel.

EASE ON HILLS

Main road hills meant nothing to the Elite. Our "test" hill has an average gradient of 1 in 12 and is as steep as 1 in 6 at one point. The Elite climbed this in third gear and was still accelerating at 70 mph when the top was reached. We have only done better than this in a sports car with almost three times the capacity—an XK 150.

There was one moment during the test which was a little disconcerting, but was not due to any fault of the car. After a burst of speed in excess of 100 mph a slight tapping noise was noticed which appeared to come from the nearside front wheel. When speed was reduced to 90 mph, the noise stopped. However, after taking a roundabout at what is quite a normal speed for an Elite, the tapping noise suddenly reappeared as a loud banging, obviously occurring with each revolution of the wheel. A halt was called immediately, and it only took a brief inspection to see that the nearside Firestone Nylon Sports type tyre had started to throw its tread. It was coming off, not in large pieces, but in small thin strands 3 or 4 inches long. At the time, it was felt that this was possibly a faulty cover, until the same thing happened to the offside tyre during maximum speed testing. It is to be hoped that the tyre manufacturers are already investigating this fault since it could be dangerous.

Since the Elite was first introduced, several criticisms have been levelled at its rather austere internal finish considering its total cost of almost £2000. Whilst we agree that it is a pity it lacks such amenities as an ash tray and sun vizors, and the matt black facia panel may appear basic, perhaps, it must be emphasized that the high cost is primarily due to its unique form of construction coupled with the very expensive development programme necessary to produce a car with these qualities of performance.

In the Elite, the manufacturers have been successful in combining five attributes—high performance, good handling, fuel economy, passenger comfort and striking appearance. They can therefore be excused one or two shortcomings, which would be dismissed in any case by most prospective owners. Of these, apart from those already mentioned, we can report that neither the nearside door nor the luggage boot would lock, and neither the boot lid nor the bonnet top boasted a supporting stay.

To sum up then, we can report that the Lotus Elite is all that it has been cracked-up to be. A car for the connoisseur, possessing everything (or almost everything) that is looked for by the real sports car enthusiast.

PERFORMANCE

Acceleration through the gears:—
0— 40 mph	5.2 sec
0— 50 mph	7.8 sec
0— 60 mph	10.2 sec
0— 70 mph	14.8 sec
0— 80 mph	18.8 sec
0— 90 mph	23.6 sec
0—100 mph	34.8 sec

Standing quarter mile 18.0 sec.
Maximum speed—110 mph.
Maximum in 3rd gear—88 mph.

The doors are trimmed in leather to match the seats, and the floor is covered with deep pile carpet. There is also adequate space for stowage of small articles in the door pockets and deep parcels shelf under the facia.

There is space for a reasonable amount of luggage at the rear of the car. The spare wheel is stored in the covered bulge at the right, accessible from within the car, and the expanding jack visible at the top left corner, is clipped on to the battery cover.

LOTUS ELITE

A FIRST LONG LOOK

By JACK GLASSON

THE arrival of SPORTS CAR WORLD at the Jolly residence in Adelaide's ritziest suburb was quite a blow to the ego. Carelessly we swung the nose of our modest Wolseley 4/44 into the backyard, stopped — and hastily reversed. There stood the Lotus Elite, a Lotus Fifteen, a Jaguar XK150, a Mercedes 220 Coupe and a couple of Holdens!

Having concealed the Wolseley safely in a side street, we put out some subtle feelers at the tradesman's entrance. Enquiry revealed, sure enough, that all these cars belonged to two of the Jolly family. The Elite, the Lotus Fifteen, the Jaguar XK150 and one Holden came under Derek Jolly's personal crest. His mother, a most charming woman, was responsible for topping up the fuel tanks of the rest.

It would have been possible to put in many a happy hour drooling over two thirds of the collection, yet we were commissioned to give impressions of the Elite. Accordingly, keeping a stiff upper lip, we centred our full attention thereon.

First impressions are that British designer Colin Chapman's newest dreamchild is low, slinky and mean. Height from the bottom of the Firestone semi-racing tyres to the top of the roof is just 3 ft. 10 in, minimum ground clearance 7 in. The two-seater, two-door integral chassis-body shell offers exceptional rigidity and strength from a virtually frameless structure of glass-reinforced epoxide and polyester resin. Externally, the body is well proportioned. Outside fit and finish is above reproach.

Front wheel location is, of course, independent by transverse wishbones incorporating an anti-roll bar. Springing is by combined coil spring/damper units operating through a single attachment point at each end. Rear suspension is again independent, using the well tried and completely original Chapman strut system incorporating combined coil spring-damper units and double articulated driveshafts, which also give lateral location. With this system a certain amount of camber change occurs with increase in load, making for good handling characteristics under most conditions.

Hydraulically operated 9½ in disc brakes are fitted, outboard at the front and inboard behind. A vertically-mounted hand brake to the left of the driver's seat operates through cables to the rear calipers.

Steering is lightweight rack and pinion. It looks suspiciously BMC.

The engine is our old friend the Coventry Climax. Four cylinders, single ohc and 1220 cc. Maximum output is 75 bhp at 6100 rpm on 8.5 to 1 compression. The engine is water cooled. It has a steel crankshaft, fully counterweighted with a big overlap between crankpins and main journals, all carried in three main bearings of the lead-bronze, steel-backed thin wall type. Aluminium pistons have plated top rings fitted. Connecting rods are split diagonally. Big end bearings are lead-bronze shells. The cylinder head is heat treated aluminium, with valves of XB steel on shrunk-in Austenitic cast iron seatings. There is a chain drive from jackshaft to camshaft, with the latter operating the valves direct through piston-type cast iron tappets. A normal spur gear oil pump with built-in relief valve pushes through a renewable-element, full flow filter. Surprisingly enough only a single carburettor is fitted — a 1¾ in horizontal SU.

A single-plate 8 in dry clutch teams with a four-speed gearbox (MGA) giving ratios of 3.635, 2.215, 1.372 and 1:1. The hypoid final drive, which once again looks like a BMC component, is arranged to take a multiplicity of ratios.

Standard: 4.55:1.
Optional: 5.125:1, 4.89:1, 4.55:1, 4.22:1, 3.73:1.

The radiator is fully ducted. It incorporates an integral header tank. The Elite's cooling fan setup is interesting. Normally a fan has to be driven from the engine through a belt. That, of course, consumes valuable horsepower. In this instance the fan is mounted *in front* of the radiator. It is thermostatically controlled so that when the temperature reaches 90 deg the electric current stops and the fan is stationary again.

The fuel tank holds a whopping nine gallons. An AC pump delivers petrol from tank to carburettor.

The electrical system is comprehensive. Ignition is by coil and distributor with centrifugal advance and retard. Standard equipment includes recessed Lucas Le Mans 7 in headlamps and separate side lamps. Flasher indicator units are grouped together with twin stop lights. The instrument lighting has a brightness control. The windscreen wiper is an electric two-speed type.

Instruments include a 4 in tachometer (0 to 8000 rpm), 4 in speedometer (0-140 mph), oil pressure gauge, petrol gauge, water temperature gauge and ammeter. All have white figures on black dials.

Knock-on 15 in wire wheels take

£2800

4.90 by 15 special high-performance Firestone tyres. Those fitted to the test car were the semi-racing pattern. The spare wheel is mounted back of the cockpit. It lies flat, housed by the rear parcel shelf. There is provision for two spare wheels, although only one is supplied. Car weight is 10¾ cwt.

The Elite's vital statistics are: wheelbase, 7 ft 4 in; front track, 3 ft 11 in; rear track, 3 ft 11 in; overall length, 12 ft; overall width, 4 ft 10 in; height to roof, 3 ft 10 in; minimum ground clearance, 7 in; weight (less fuel), 10¾ cwt.

Having disposed of technical details, let's move on to a closer inspection. The glass fibre bodywork is without a blemish. Its white paintwork glistens in the sun. The big wraparound windscreen and rear window should certainly afford good vision — but here the first of several annoying features rears its ugly head. The doors are *not* provided with wind-up windows. Instead they have detachable side panels of glass, plus hinged quarter windows for ventilation. Although the side panels are quickly detachable and can be stored in capacious door pockets, the fact remains it is impossible to give a hand signal while they are in.

Entrance to the driving seat is good, although the Elite can be an awkward car to get out of.

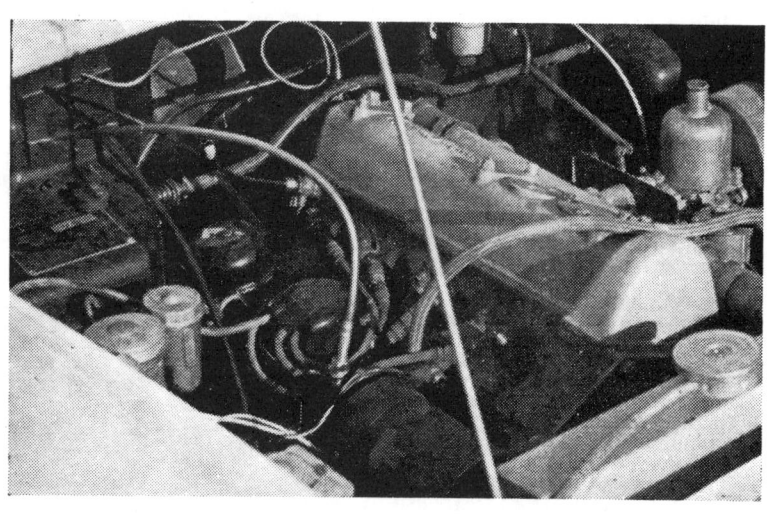

Tilted, single cam Coventry Climax turns 75 brake at 6100, uses only one SU. Bonded-in scuttle and valance bracing panels guard against flexing in fibre shell.

Sharp's the word. Sweet lines are Elite's alone, combine beauty with wind-chiselling functionalism. Single big fault is in side windows, which don't wind.

Properly set up behind the wheel, you really feel this is a thoroughbred sports car. Close, form-fitting seats (the most comfortable we have yet tried) hug you firmly. You know in these you will stay put in the tightest corners. The pendant foot pedals are exactly right, but dipper switch and windscreen washer button are on the floor very close together. Conceivably, you can get a screen full of water while the other bloke goes right on copping the chrome-blistering impact of your Le Mans high beams!

The stubby gear lever falls right to hand. There's ample room to swing the three-spoke duralumin wood-rim wheel. Headroom is not very great, although the inside of the roof has no true lining — only a type of felt sprayed on to the fibre.

Behind the driver, at each side of the rear parcel shelf, are most interesting humps known to Elite fanciers as the car's Marilyn Monroes. In actual fact they house the rear suspension coil-damper units.

Space for both driver and passenger is generous. The driving position itself is magnificent. The seat is inclined slightly backward, making for a good straight-arm stance.

Time to try. Turning the ignition key to the extreme right engaged the self-starter. The cold engine started immediately. Gingerly we floored the clutch and laid a hand on the 4 in high, centrally-mounted gear lever to select reverse. Moments later we were still trying to find the same gear. Eventually we discovered that because of its short length there is a complete lack of leverage with the gearstick. It's almost a two-handed job to select reverse in a new box. Perversely enough, the forward gears are sweet and smooth. They require only wrist-action for quick, clean changing.

Steering is light, direct and quite effortless. At virtually any speed the car can be controlled with but two fingers on the wheel. Although we in the South had not had the chance before to test a car fitted with disc brakes, reports of their stopping power had reached us. We were not disappointed. The Elite's brakes literally lift you off the seat, albeit at the expense of high pedal pressures.

Roadholding is superb, be it in an 80 mph curve or through a pot-holed dirt-road hairpin. The car simply stays put with a complete absence of tyre noise. It will take deliberate attempts to throw it off line nonchantly in its stride. The Elite corners quite flatly. Although the tail appears to be harshly sprung, the rear wheels do not dance in turns. Even under hard braking or fierce acceleration in a corner the car corrects immediately.

The little Lotus is a willing worker through the gears. The lusty Climax engine screams effortlessly through the whole rev range on demand. A determined operator can lay two strips of rubber on the road from a standing start. Stopwatch figures collated by the Royal Automobile Association of South Australia (and, therefore, not guaranteed) indicate acceleration 0-50 mph in 8.6 sec and 0-60 in 11.2. Please remember test car was not really run-in. Slightly better times should be possible later. Most surprising is the little beauty's RAA fuel consumption — 55 mpg at 55 mph.

During our afternoon's road session with the Elite we found that the cockpit got rather hot and that engine noise was prevalent. We attribute both in some degree to an opening cut in the test car's firewall. Neither should apply to the same extent in other cars of the make. A glass fibre bodyshell does not dissipate heat as readily as a metal one can, and although the temperature that day was in the 80s the interior of the stationary car got rather hotter than we would have expected. On the move, of course, draughts of cool air could be directed in through quarterlight windows.

Aside from detail criticisms, we would be hard to please if the car did not impress us. Truly, it did. Graceful lines, sparkling performance, magnificent roadholding, impressive braking, and relaxed seating leave an indelible impression that has not been approached by any other sports car in our personal experience. Wholeheartedly, we recommend the Elite to the enthusiast who looks for quality with performance. In fact, even if you prefer it as a town car, the little Lotus will serve you well. It is highly manoeuvrable. The engine will pull away from as low as 15 mph in top gear without sign of stress. We place the Elite high on our list as a sports car whose racing heritage is built-in. Price is £2885 tax paid. Delivery can be given three months after the order is received. Sole Australian agent is D. E. Jolly, at 16 Robe Terrace, Medindie, SA. #

Few sports coupes offer luggage room like this. Elite's stressed skin construction allows peak utilisation of enclosed space, perhaps offsetting lack of back seat.

Dreams are made of stuff like this. Black instrument bezels, severe white-on-black lettering, liberal matt padding, ideally placed controls plus built-in Halda pilot.

BEST OF THE REST *Great Britain*

Elite is as handsome as it goes well. Beautiful glass body is at home in any company and the car has been very successful in all sportscar races.

LOTUS ELITE

The Elite is a classic example of the maxim which says that if a thing looks good, it is good. It is the most attractive small coupe that we have ever seen out of England — a country noted for stodgy design policies — and it goes. Light weight from its fiberglass body and a beautifully aerodynamic shape allows the Elite's tiny 1220cc (74.25-cubic-inch) overhead-cam Coventry Climax engine to push the car up to 110-115 mph, depending upon rear axle ratio. Acceleration to 60 mph takes only about 10 seconds and this is with the single-carburetor 75-hp engine.

Legroom is quite good; luggage space divided between small trunk and behind the seat is about average for a 94-inch-wheelbase small car. Girling disc brakes, inboard at the rear, provide more than enough retarding to stop the Elite's 1205 pounds.

Technical facts fail to tell the story of the Elite's charm. Quick 2½-turn steering allows instant maneuverability around the tightest bends at practically any speed without undue effort. Independent rear suspension insures all the power getting to the road. The ride is firm over bumpy roads but cruising at 80-90 mph on smooth highways is completely comfortable. In short, the Elite is a great small car design.

Large rear window gives almost open-car visibility. Because be snapped in and out. Vent panes control air circulation. the doors are curved, windows cannot be rolled but must

ROAD TEST/14-61
LOTUS ELITE *How Grand Can Touring Be?*

PHOTOS BY PAT BROLLIER

ONCE IN A GREAT WHILE, no matter how many cars one has an opportunity to live in close harmony with, along comes an automobile that for a variety of reasons — not all of which are entirely rational — comes close to subjective perfection.

We have yet to see a car that is entirely wrong and — in the cold light of pure objectivity — one that is entirely right. But on rare occasion we fall in with an automobile that fits one's likes, aptitudes, abilities and physical shape so well that the temptation to disappear to some remote and untraceable section of the globe grows to almost irresistible proportions. Such a car need not be extraordinarily fast, though in our case most have been, nor need it be outrageously expensive, though perfection or near-perfection does not come cheaply.

The latest in our list of nominations for this particular group of machines is Mr. Colin Chapman's delightful confection for those who would tour grandly, the Lotus Elite. While it is hardly a new car, having been brought forth more than three years ago, few have come into public view until quite recently. The reasons for the delay are nobody's business but Mr. Chapman's; it is enough that the machine is now reasonably available to those who would search out a Lotus distributor and for some — though certainly not everyone — it will be a rewarding search indeed.

To begin with, despite a certain simplicity, the Elite is one of the most technically advanced automobiles on the road. In concept is it derived directly from the front engined Grand Prix car of 1958 and '59, utilizing virtually the same suspension system as that machine. Another unique feature and one that hasn't been emphasized strongly enough even now is the unit body-chassis structure that is constructed almost entirely of glass fiber reinforced resin. So rigid is this structure that only two small steel subframes are needed. One of these, molded into the glass, supports the engine and front suspension while the other forms a sort of cage around the cockpit area, tying the whole structure together. This latter also has hoops that come up through the windshield posts and over the windshield itself to form a roll bar. A smaller hoop comes up and over the rear of the cockpit. Stiffening at load points throughout the body-chassis construction is handled by varying the number of sheets of glass fiber, thus thickening load-bearing areas. The result is extreme lightness coupled with immense strength. To those accustomed to considering strength in terms of structures as the Brooklyn Bridge such construction might seem flimsy but doubters need only to have seen two examples of the Elite which had been thoroughly rolled and bounced in racing mishaps. In one case the car flipped and rolled several times. In both cases the cars were towable and in neither case was there internal cockpit damage, the majority of the injury being in splintered and abraded bodywork. Both drivers walked away from the wrecks. In short, the Elite is fully as strong as it needs to be.

Weighing in at only 1512 lbs., and with the suspension system of a Grand Prix car, one would expect handling of a high order. Just how the order of reality is comes as a distinct shock. At low speeds in the city the handling qualities do not manifest themselves to any great extent; in fact things seem just a bit stiff which might mislead one into worrying about a tendency to skitter at higher speeds. Not so. The higher the speedometer needle travels the tighter the car seems to cling to the road surface and the smoother the ride becomes. Adhesion to the road gets to a point that becomes just a little eerie. This is not a car to be blithely tossed about dirt-track style but one which must be *driven* through bends and turns.

(continued on following page)

though there seems to be an excessive amount of body lean on hard cornering, the effect on the driver is negligible; one rdly notices it. The major attribute of the Elite is that it sticks fast to the road at both ends and steer characteristic dead neutral. Fast turns can be taken as much as 20 mph faster than usual with a weird feeling of complete security.

Passenger's side of the cockpit is roomy and comfortable with all sorts of leg space. Secure, high backed seats will hold one in tightest turns. Parcel shelf is directly ahead.

The driver's office is one of the best we've seen with every control comfortably reachable. The instruments are instantly readable, well-lit. Hip room is on the slight side. Reduce!

Windows (far left) are of the snap-in variety and clip easily in place. Trunk space is generous for car the size of this one. Annoying point is that there is no trunk-lid hold rod.

Small engine, big punch. Docile-looking, single carbureted Coventry Climax four-banger produces 75 horses from 74 ins. Range is fantastic and power sufficient to produce 120 mph.

LOTUS ELITE

Steering is light and with almost nil return action, the sort of thing designers of power steering probably had in mind but failed to achieve. This light, positive control coupled with the uncanny ability to stick to the road leads one eventually to experimenting with turns. On a trip between Los Angeles and Las Vegas we found ourselves continually going through turns that could have been taken with normally good handling equipment at, say, 55 or 60 mph but which in the Lotus were so safely and smoothly taken at 20 mph more that at times we felt slightly foolish. Since the trip was being made to attend the Las Vegas races the road was pretty well filled with other sports cars and on occasion their drivers took exception to being passed in what must have seemed to them an unnecessarily blithe manner. Their efforts to make up for the seeming insult could at will, with only a slight additional pressure of the right foot be brought to nought. As mentioned — uncanny.

Power for these shenanigans and others such as may be seen by glancing at the acceleration graph, comes from a docile but extremely willing version of that ubiquitous firepump, the SOHC Coventry Climax — 74 cubic inches and 75 horsepower's worth. The number of horses living under the hood do not, however, spell out the whole story. The spelling is R-A-N-G-E. While the horsepower peak is delivered at a healthy 6100 rpm, the torque peak is way down there at 3400 rpm with the result that at just about any speed a poke at the throttle produces immediate results. Despite its small size and relatively hot performance you do not row this car around with the gearbox. The box is there to be used, and it should be, but it isn't necessary to be forever waggling the stick around like the handle of a butter churn. On our tests at Riverside Raceway, normally a three gear course, it was utterly unnecessary to use anything but third and high with third gear being used only where most cars require the use of second. This ability to accelerate from just about any point on the dial enabled us at one point to play fun and games with a gentleman in a Lincoln who formed the annoying habit of passing and then slowing down below our normal cruising speed on a straight stretch of four-lane road in Nevada. We had passed him at a steady clip and then moved into the right lane. He boomed around and then slowed down. We passed him again and the act was repeated. The third time we waited until the behemoth was alongside and then poked the throttle. The Elite squirted ahead like a squeezed watermelon seed and we proceeded on down the pike at something in hand over the century mark for a sufficient number of minutes to end the man's attempts at free-lance traffic regulation.

Strangely enough all this sizzle is accompanied by economy of fuel consumption bordering on the fantastic. During performance testing the Stage I Elite showed an average of 30 miles to the gallon despite continuous laps at racing speed, acceleration runs and top speed, full-throttle tests. In city driving the average jumped to 38 mpg and on the Las Vegas trip where speeds were fairly steady, though with the several high speed bursts noted above, the figure was a near-unbelievable 43 miles to the gallon. No attention was paid at any time to feather-footing for economy reasons though some lightfoot work was necessary just to avoid tangling with the law. While part of this economy is due to the small size of the engine with its lone carburetor, most of it can be pegged to the light weight and aerodynamic shape of the car.

Firing up the Elite is an instantaneous business involving turning the key to the right at which time the Climax bursts busily into an 800 rpm idle. Given a minute or two to warm up, throttle response is of the right-now variety with the tach needle literally whizzing across the face of the instrument. Busy-ness is accompanied by considerable buzzy-ness

ROAD TEST/14-61
TEST DATA

VEHICLE .... Lotus   MODEL .... Elite
PRICE ((as tested) ..... $5741   Options .... Stage 2 & 3 power kit,
                                 CR gears, Radio and Heater

ENGINE:
Type: .... 4 cyl., 4 cycle, in-line, water-cooled
Head: .... Aluminum
Valves: .... SOHC, inverted tappets
Max. bhp .... 75 @ 6100 rpms
Max. Torque .... 72 lbs. ft. @ 3400 rpms
Bore .... 3.0 in. 72.2 mm.
Stroke .... 2.62 in. 66.6 mm.
Displacement .... 74 cu. in. 1216 cc.
Compression Ratio .... 10.0 to 1
Induction System: .... Single SU carburetor
Exhaust System: .... 4 branch iron manifold
Electrical System: .... 12V Lucas

CLUTCH: .... Single-disc, dry   DIFFERENTIAL:
Diameter: .... 8 in.   Ratio: 4.55 to 1. Optional 3.7, 4.2, 4.87
Actuation: .... hydraulic   Drive Axles (type): Open, independent

TRANSMISSION: STEERING: .... Rack & Pinion
Ratios: 1st .... 3.67 to 1   Turns Lock to Lock: 2½
        2nd .... 2.50 to 1   Turn Circle: .... 34 ft.
        3rd .... 1.34 to 1   BRAKES: .... Girling discs — inboard
        4th .... 1.0 to 1   rear, outboard front
                                       Drum or Disc Diameter ...... 9.5 in.
                                       Swept Area .... 359.36 sq. in.

CHASSIS:
Frame: .... Steel tube molded in Fiberglas
Body: .... Fiberglas, semi-integral
Front Suspension: .... Unequal A-arms, coil-shocks
Rear Suspension: .... Lotus strut-type, coil-shocks
Tire Size & Type: .... 4.80 x 15 Michelin X

WEIGHTS AND MEASURES
Wheelbase: .... 88 in.   Ground Clearance .... 6.5 in.
Front Track: .... 47 in.   Curb Weight .... 1512 lbs.
Rear Track: .... 47 in.   Test Weight .... 1735 lbs.
Overall Height ..... 46 in.   Crankcase .... 4 qts.
Overall Width ...... 58 in.   Cooling System .... 10 qts.
Overall Length ..... 130 in.   Gas Tank .... 7.0 gals.

PERFORMANCE
0-30 .... 3.1 sec.   0-70 .... 13.8 sec.
0-40 .... 5.2 sec.   0-80 .... 18 sec.
0-50 .... 7.1 sec.   0-90 .... 22.2 sec.
0-60 .... 10.6 sec.   0-100 .... 29.8 sec.
Standing ¼ mile 19 sec. @ 82 mph. Top Speed (av. two-way run) 119 mph
(@ 7000 rpm)
Speed Error   30   40   50   60   70   80   90
Actual   29   39   48   57   67   77
Fuel Consumption Test: ..... 30 mpg
Average .... 34 mpg
Trip .... 42 mpg   Speed Ranges in gears:
Recommended Shift Points   1st .... 0 to 30 mph
Max. 1st .... 30 mph   2nd .... 5 to 50 mph
Max. 2nd .... 50 mph   3rd .... 20 to 83 mph
Max. 3rd .... 83 mph   4th .... 30 to 119 mph
RPM Red-line .... 6500 rpm
Brake Test: .... 75 Average % G, over 10 stops
Nil Fade encountered on 10th stop.

REFERENCE FACTORS
Bhp per cubic inch .... .98
Lbs. per bhp 18.9
Piston Speed @ Peak rpm .... 2664 ft./min.
Sq. in. swept brake area per lb. 253

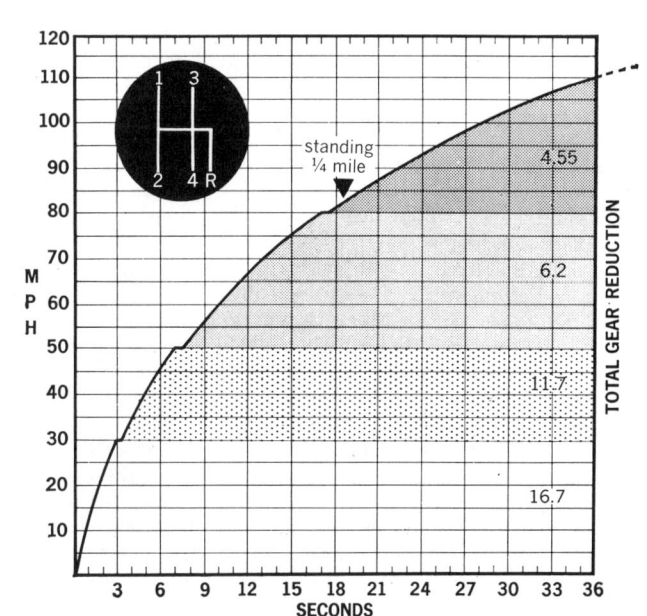

LOTUS ELITE

in this particular case but it isn't annoying; it's just there to remind you that a lot of small parts are doing their jobs efficiently and rapidly.

The clutch, hydraulically operated, is feather-light to the touch and very smooth on engagement. It seems almost too soft until it is realized that the bite is solid and firm with every engagement after a gear change biting home solidly with no slip or other nonsense associated with soft-pedal clutches. The gearbox is a standard MGA unit complete with the usual BMC gearing. The close ratio gears used in the MGA competition box can be had for an additional $80 in a package. If you want them installed it is an additional $90 labor. This is simply due to the fact that the CR gears are not part of the regular BMC line and somebody has to stuff them into the case after it leaves the assembly line. Our test car had the normal gearing and, as mentioned, a higher, closer ratio second gear wasn't missed although it is quite possible that it could come in handy on tighter courses.

As hinted at earlier, acceleration feel is strong in spades. It isn't of the slam in the back variety but a sort of incredibly smooth increasing swiftness that seems to build and build as long as one has stomach enough to hold down the throttle. Cruising speed in top gear can be anything the law allows from 55 or 60 on up to 90-plus with the exception of one spot around 70, or in the neighborhood of 4400 to 4700 rpm, at which point there is a strong vibration. We are told this is endemic with the breed and that much time and money has been expended on finding the cause and cure. It isn't damaging but it *is* annoying and one soon learns to drive around that point — in fact it is one of the very few negative attributes of the car. The others: no support struts for the hood and rear deck lid, an inaccessible battery, flimsy bumpers and weak, soft rubber rear exhaust supports. That's about the list unless one wants to include a rather high interior noise factor (but no wind noise — *none*) which didn't bother us particularly but might annoy those with more delicate ears; to be truthful we sort of enjoyed it. Oh, yes — there was one other thing. With a huge drive-shaft tunnel running between the seats which had a nice flat surface on which one could place cigarette packages and the like there was ample acreage for an ashtray a-la Healey but no ashtray was present — a small touch but in a car that is as expensive and well-thought out as this one, an irksome point.

Getting back to plus factors (pull up a chair) as a car goes, so should it stop. This one goes and stops with equal dexterity. The stopping power is provided by Girling 9½-inch discs all the way around, outboard on the front and inboard on the rear, eliminating brake torque on the rear hubs. During tests and afterward during hard racing laps around Riverside there was no tendency to fade or to pedal loss.

Steering as mentioned is feather-light and positive. Tracking is so good that it becomes a hands-off proposition at any speed on a smooth road. The car is instantly responsive to any wheel movement but not so much so that over-correction is ever a problem. Insofar as we had guts enough to check it, steer characteristic is dead neutral with both ends sticking to the road as though they were built into it. As the photos show there is more than a little lean on hard cornering but it is almost completely imperceptible to the driver.

The ride, as mentioned, seems just a bit stiff on rough streets at low speeds with the small bumps making themselves felt but as speeds increase this is all ironed out with little perceptible difference being noticed between bumpy roads and smooth ones either in control or ride. Despite the small size of the car, the problem of getting in or out is not a large one unless one is more than six feet tall or one's lady passenger is wearing a tight skirt. Even for a six-footer, once he's inside, there is ample room. The seats are firm and snug with all the lateral support anyone could possibly want and plenty of arm and shoulder maneuvering room. The driving position is typically Lotus legs-out-ahead and with a vertical wheel placed perfectly for arms-out handling. The instruments are of an order to delight the heart of the purest of the pure sports car enthusiasts, their black faces with white numerals showing just about anything one would want to know. Placement is perfect with no blind spots and the lighting is an unobtrusive blue that picks out the numbers without blinding the driver. The small controls are where they belong, easy to memorize and to operate. Typical GT practice is the combination horn and light flicker switch. On a long stalk near the wheel a movement downward blows the horn and a flick upward turns on the high beam lights whether the regular light switch is on or off.

One of the objections to a small car for touring is usually that there isn't enough space to store luggage and other impedimenta. Not so with the Elite. The trunk looks small but it will handle one standard suitcase with a camera bag (large), loose clothes on hangers and an untold amount of softly packed goods. On the shelf over the spare tire behind the seats there is room for a small suitcase. In each door there is a capacious pocket that will carry anything from books to laundry. In addition there is a spacious package shelf under the dash. If necessary and if the chattels are properly stowed one should be able to lug one to two weeks' worth of supplies along on a trip in this car.

The windows in the doors are of the snap-in variety first seen in the gull-wing Mercedes 300 SL and can be snapped into place almost as fast as a window could be wound up. Ventilation control is handled by windwing windows pivoted in the forward part of the doors. So sensitive are they to opening position that virtually any degree of air venting from a blast to a zephyr can be had. A nice touch here is that they are equipped with small diagonal gutters that prevent dripping from the inside corners in a rain. Visibility is excellent all the way around except for one small blind spot over the driver's shoulder to the right in the case of a left-hand drive car and to the left in the case of a right-hand drive. This can easily be eliminated with accessory outside mirrors.

At $5714 the Elite is not an inexpensive car — in fact for its size it's a pretty expensive one. Further, at first glance it doesn't look like that kind of money and for that reason it isn't the sort of car that can be "worn" like a diamond stickpin as a status symbol — but then again it wasn't meant to be. Nor is it a loud, fierce, impressive looking car. It is, rather, a car to *worry* the man with a loud, fierce, impressive looking car. In short, the Lotus Elite is not everyman's sports car (should there even be such an item) but rather it is an esoteric machine built for those who demand the ultimate in grace, performance and, above all, handling in a unique Grand Touring car. For the man to whom these things are important, or to whom production class racing in such a machine is important, the Lotus Elite is worth the price of admission. There may be bigger cars and there may be more luxurious ones but there are few that will give more than the Elite in terms of sheer sensuous driving pleasure. — *John Christy*

JOHN BOLSTER TESTS

The Lotus Elite—"LOV 1"

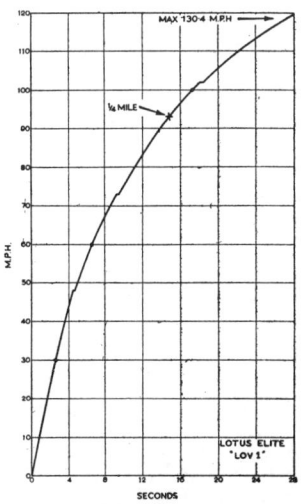

ACCELERATION GRAPH

LAST May, AUTOSPORT tested a standard Lotus Elite which was submitted by the manufacturers. Good though the performance of this car was, it obviously bore no relationship to the urge displayed by the ones that are raced by "the boys". Therefore, to complete the Lotus picture, I have recently borrowed the most successful racing Elite in the country.

Graham Warner's "LOV 1" is well known to all those who visit the circuits. I collected the machine exactly "as raced" —indeed, it had not been touched since notching up yet another first place at the Hatch. Naturally, it had a "low cog" in for that circuit, but as the tuned Stage III Coventry Climax engine is perfectly safe at 8,000 r.p.m., this was no great hardship. With a Le Mans ratio buttoned into the final drive, there is no doubt that this Elite would be a 140 m.p.h. car. With the Brands Hatch ratio fitted, however, a rapid calculation from the tyre size, plus the centrifugal expansion factor, showed that one could enter the 8,000 r.p.m. band as the speed approached 130 m.p.h. Nevertheless, the loss of that ultimate 10 m.p.h. was counteracted by the more brilliant acceleration that the "low cog" provided, and so I was well content to take the car as it stood.

The body is somewhat short of interior trim, but sponge rubber padding is placed at strategic points. The suspension is standard, except that the dampers are adjustable in front. Light alloy disc brake calipers are fitted.

Perhaps the most important non-standard feature is the ZF gearbox. This has synchromesh on all four speeds, and the ratios are sufficiently close to keep the willing engine well up in the rev. range. The synchronized bottom gear is high enough to be really useful for sharp corners.

The 1,216 c.c. Coventry Climax engine is modified to Stage III tune by Cosworth Engineering, Ltd., and develops 98 b.h.p. at 7,200 r.p.m. As the standard Elite gives 75 b.h.p. at 6,100 r.p.m., the source of the extra performance is at once apparent. A four-branch exhaust system has been developed for maximum power production on the circuits, and little thought has been given to silencing.

When one first drives "LOV 1" on the road, one is extremely conscious that the somewhat shattering exhaust note may lead to trouble. However, the maximum sound output occurs at around 4,000 r.p.m., and if one keeps well below that figure a trip through London is perfectly practical. In the open country, I kept well above the 4,000 r.p.m. band for the most part, and the delightful singing note then emitted was by no means raucous.

Although the chassis had not been modified, the "feel" of this car differed a good deal from the standard model previously sampled. The stability at speeds well above 100 m.p.h. was really excellent, and no car could be easier to handle at such velocities. At comparatively low speeds, the directional stability was less marked, a very slight tendency to wander being noted on certain occasions. The light, quick steering gave superb control, at the expense of some kick-back over bumps.

No car could be easier to take off the mark, and this all-independent machine demanded none of the driving skill that a rigid rear axle would require. At something like 5,000 r.p.m., I simply took my foot off the clutch pedal and disappeared—but quickly! At a crowded massed start, it must be comforting to know that a little over-enthusiasm will not result in hopelessly excessive wheelspin. The gears may be changed as fast as the hand can move, and the clutch takes hold at once.

The acceleration is of an entirely different order from that of any production sports model, and on the road the most powerful cars are simply left standing. The pleasure of handling the car is greatly enhanced by the smoothness of the engine. Even at 8,000 r.p.m. there is not a tremor, and this speed is reached in top gear on any short straight. There is absolutely no sense of strain, the power unit appearing to enjoy the fun as much as the driver. It does not get hot, use oil, or run on when switched off.

The aerodynamic shape must be exceptionally efficient. This results in quite outstanding acceleration in the upper ranges. The astonishing liveliness between 80 and 110 m.p.h. is something which much larger competition cars find difficult to emulate. As the steering and the brakes are well up to the performance, it is easy and safe to achieve very high average speeds, even on fairly crowded roads.

No sign of temperament was shown during the whole of the test. The engine is flexible, even by touring car standards, and it will idle quietly in traffic without any danger of wetting its sparking plugs. Apart from the somewhat healthy exhaust note, one cannot fault the manners of this Elite. The fixed-jet SU carburetters need choking with a couple of lumps of rag on a cold morning, but at all other times starting is instantaneous without opening the bonnet.

A rather remarkable feature of the performance of "LOV 1" can be appreciated if one looks back at my earlier road tests of "works" Lotus sports-racing cars. Within a remarkably short period, a comfortable closed car has been produced which can better the figures of quite recent "racers". This is progress indeed. "LOV 1" is not habitually used on the road now, because in Stage III tune her voice is too loud. Nevertheless, for a few days she made me the master of the Queen's highway, and I am grateful to Graham Warner for lending me the car that has brought him so many successes.

SPECIFICATION AND PERFORMANCE DATA

Engine: Four cylinders 76.2 mm. x 66.6 mm. (1,216 c.c.). Single overhead camshaft directly operating valves through inverted piston tappets. Light alloy head and block. 98 b.h.p. at 7,200 r.p.m. Twin SU carburetters. Lucas coil and distributor.

Transmission: Single dry-plate clutch. ZF four-speed gearbox with short central lever and synchromesh on all gears. Ratios 4.9, 6.02, 8.38 and 12.40 to 1. Chassis-mounted hypoid final drive unit.

Chassis: Plastic integral body and chassis. Independent four-wheel suspension by helical springs and telescopic dampers; front suspension by wishbones, anti-roll bar and rack and pinion steering; strut-type rear suspension. Girling disc brakes, inboard at rear. Knock-on wire wheels fitted Dunlop R5 tyres (front) 4.50 x 15 ins. (rear) 5.00 x 15 ins.

Equipment: Twelve-volt lighting and starting. Speedometer, electronic rev. counter, oil pressure and water temperature gauges, ammeter.

Dimensions: Wheelbase, 7 ft. 4 ins.; track, 3 ft. 11 ins.; overall length, 12 ft. 6 ins.; width, 4 ft. 10 ins.

Performance: Maximum speed, 130.4 m.p.h. Speeds in gears: 3rd, 102 m.p.h.; 2nd, 73 m.p.h.; 1st, 48 m.p.h. Standing quarter-mile, 15.1 secs. Acceleration: 0-30 m.p.h., 2.6 secs.; 0-60 m.p.h., 6.6 secs.; 0-80 m.p.h., 13.8 secs.; 0-100 m.p.h., 17.2 secs.

Fuel Consumption: Driven hard, 20 m.p.g. (approx.).

JOHN BOLSTER TESTS

The "Do-It-Yourself" LOTUS ELITE

Some years ago, I led a deputation which tackled the Government on the question of purchase tax for home-built cars. I enjoy the rough and tumble of politics, and our efforts received considerable notice in the House, as Hansard testified. A member of the deputation was Colin Chapman, and he has now gone a stage further by persuading the gentlemen of H.M. Customs and Excise to sanction a do-it-yourself Lotus Elite.

This, of course, is an extremely important step forward for the motoring enthusiast. Most of the younger generation are now mechanically minded, and under suitable conditions they are quite capable of building a complete car from a set of components. To earn the purchase tax concession, a car must genuinely be constructed by an amateur without professional assistance. A man who has acquired the necessary know-how, and is willing to put in plenty of hard work, can become the owner of a new Elite for £1,299 instead of £2,006.

My colleague, Patrick McNally, and I have recently been sharing an Elite that has been built from the standard component set. It must be understood that this car was in normal fast-touring tune and was not set up to tackle "LOV 1" or "DAD 10". Thus, the Coventry Climax engine was in Stage I tune, though with twin carburetters, and a very effective double exhaust system rendered the machine entirely suitable for London. The M.G.A gearbox also had the standard ratios, which would be rather wide for tackling the circuits. Nevertheless, the basic car was identical with the "racers" and the engine and gearbox could be converted at any time if competitions were in mind.

The basis of the Elite is a combined body and chassis in fibreglass, there being no conventional metal frame. A tubular frame with metal body panels might be just as light, but sound insulation—always a problem in a small closed car with a high-efficiency engine—would present serious difficulties and could well require the addition of heavy damping material. The two-door two-seater body has a useful luggage boot and there is also some storage space on and around the spare wheel cover behind the seats.

The Coventry Climax engine is mounted at the front on rubber, a static propeller shaft taking the drive to a combined hypoid unit and differential, which carries the inboard disc brakes. The rear suspension, of the strut type, projects into the rear compartment, where the helical springs are covered by a pair of cylindrical jackets of unusual and misleading shape. "No, they're not bottles, offisher!" The front suspension is by wishbones, helical springs and an anti-roll bar, with rack and pinion steering.

As the leather-covered bucket seats are carried very low, a deep propeller-shaft tunnel is employed. The short, central gear lever projects from the top of this, but the handbrake, rather surprisingly, is an umbrella handle under the instrument panel. A hinged panel gives reasonable access to the power unit, but there is nothing to support it in the open position.

The engine is the well-known Coventry Climax FWE light alloy unit. It has wet liners and a single, chain-driven overhead camshaft. With a bore and stroke of 76.2 mm. x 66.6 mm. (1,216 c.c.) it develops 80 b.h.p. at 6,000 r.p.m. on a compression ratio of 10 to 1. The single dry-plate clutch is hydraulically operated and the gearbox, which is in unit with the engine, has synchromesh on the upper three ratios.

When I took over the car, I was at once impressed by the seating position, the location of the wood-rimmed steering wheel, and the pedals, which are ideal for heel-and-toe. The enging starts easily on the switch, and it is at once obvious that the silencing arrangements have been greatly improved. The straight-cut spur gears of first speed howl cheerfully, but the other gears are not obtrusive. The final-drive unit hums on the overrun but is commendably quiet on acceleration, and road noise has been reduced.

The Lotus Elite is an ideal fast-touring car in almost every respect. The engine always sounds "busy" but it is the encouraging sound of willing machinery enjoying its job. It is very flexible, too, but the revs. must be kept up if a lively performance is desired. The lower gear ratios are rather widely spaced, but 80 m.p.h. may be reached on third speed, which permits some spirited overtaking.

The aerodynamic shape is very good indeed. This encourages the car to go on accelerating past the 100 m.p.h. mark in a way which is reminiscent of a 3-litre machine. The rev.-counter enters the red section at about 110 m.p.h. which may be regarded as the normal maximum, but 115 m.p.h. is actually available if one lets the needle encroach on the forbidden zone. In any case, the unit has an enormous safety margin in Stage I form.

The getaway from a standing start is truly excellent on all surfaces and independent rear suspension really pays dividends under these conditions. The rear suspension has recently been improved, a bottom wishbone having been added, and the cornering power is very high indeed, with absolutely no vices. On dry roads, the speed at which one can flash through corners is something at which to marvel. On wet roads, there is not much "feel", and it takes practice before corners can be taken on the limit with absolute certainty. Having got the message one can continue to drive with considerable verve even on slippery surfaces.

Fairly firm pressure on the pedal is demanded by the disc brakes, and an emergency stop really exercises the leg muscles. Nevertheless, the brakes are extremely powerful and the superb road-holding allows them to be used fiercely in safety. The handbrake is, I am afraid, the only part of the car which one cannot avoid criticizing. The clutch grips well and can cope with racing gear-changes. The gearbox of the test car was rather stiff, first and particularly reverse being difficult gears to engage on occasion, but a greater mileage would no doubt bring about an improvement. The engine tended to be thirsty of oil, and this again could be due to insufficient mileage to bed in the chromium-plated piston rings.

An outstanding feature of the Elite is its petrol economy. The car will return a full 30 m.p.g. when driven hard, and during ordinary fast touring it can even better this. A simple heating system is employed which is not powerful but gives some comfort to the driver and passenger. Reflection in the windscreen was experienced unless the instrument lights were turned very low at night, and a little more power in the headlamps would not come amiss for really fast driving in the dark. The door windows have no winding handles but may be removed for ventilation, though the swivelling panels normally suffice.

As a fast touring car, both in Britain and on the Continent, the Elite has many virtues. The impeccable roadholding on all surfaces is a great asset, and the ride is always comfortable with a total lack of pitching. No rattles were noticeable on the test car, and the interior sound level, though never approaching silence, is moderate for the type of car. Owing to the unobtrusive exhaust note, the car can be driven hard without attracting the attention of the wrong people.

The Lotus Elite must be one of the safest cars on the road. As a touring sports car for everyday use, its original faults have now been rectified, and it is a well-tried and thoroughly satisfactory machine. Nobody has ever denied its superb appearance, and the necessary equipment is always available to convert it into a potent competition model. The racing car of today is the touring car of tomorrow, and this is it.

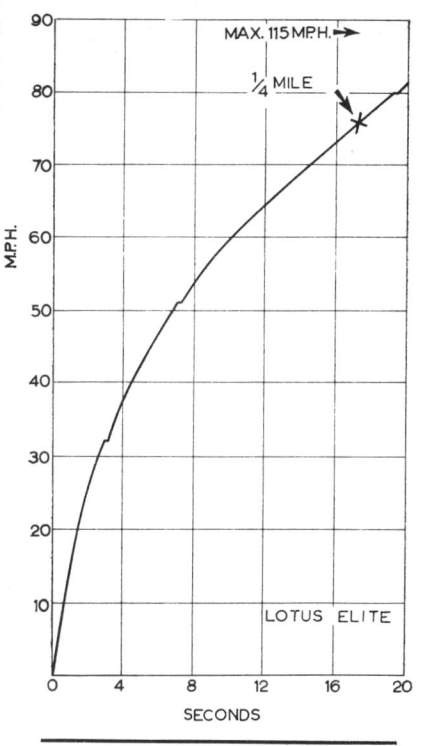

SPECIFICATION AND PERFORMANCE DATA

Car Tested: Lotus Elite sports two-seater coupé. Price £1,299 in kit form.

Engine: Four-cylinders, 76.2 mm. x 66.6 mm. (1,216 c.c.). Single overhead camshaft. Compression ratio, 10 to 1; 80 b.h.p. at 6,000 r.p.m. Twin SU carburetters. Lucas coil and distributor.

Transmission: Eight ins. single dry-plate clutch. Four-speed gearbox with central remote control and synchromesh on upper three gears. Ratios, 4.55, 6.00, 9.10 and 16.70 to 1. Static propeller shaft. Chassis-mounted hypoid unit with articulated half-shafts.

Chassis: Combined body and chassis of fibreglass. Independent front suspension with wishbones, helical springs and anti-roll bar. Rack and pinion steering. Independent rear suspension by struts, helical springs and wishbones. Telescopic dampers all round. Disc brakes on all four wheels. Centre-locking wire wheels fitted 4.80-15 ins. tyres.

Equipment: Twelve-volt lighting and starting. Speedometer, rev. counter, ammeter. Petrol, water temperature and oil pressure gauges. Heating and demisting. Windscreen wipers and washers. Flashing direction indicators.

Dimensions: Wheelbase, 7 ft. 4 ins. Track, 3 ft. 11 ins. Overall length, 12 ft. 6 ins. Width, 4 ft. 10 ins. Weight, 13 cwt. 1 qtr. (weighed with one gallon of petrol).

Performance: Maximum speed, 115 m.p.h. Speeds in gears: 3rd, 80 m.p.h.; 2nd, 51 m.p.h.; 1st, 32 m.p.h. Standing quarter mile, 17.2 secs. Acceleration, 0-30 m.p.h. 2.6 secs., 0-50 m.p.h. 6.8 secs., 0-60 m.p.h. 10 secs., 0-80 m.p.h., 19.2 secs.

Fuel Consumption: 30.5 m.p.g.

An Elite in the family

Daringly conceived ... painlessly born

By Bill
Dale-Otis

EVER since I saw my first Elite something over three years ago I have always regarded it as one of the prettiest small cars ever made. The crisp, unadorned perfection of its lines, the graceful front with its ingeniously accommodated number plate; the amazing lowness of the car, the whole happy air of being beautiful and functional, the two so intertwined it is almost impossible to say where one ends and the next begins.

It was not long after seeing the first one and reading about its unique monocoque glass fibre construction that I saw positive proof of its capabilities on the racing circuit. But however much I coveted the graceful little beast there was a major obstacle—price. Assuming one had the money, how could one justify something around £2,000 for a car that would be basically "pleasure only"? So I just watched and marvelled at the performances of the Warners, the Lestons and the Hobbs on the race track, meanwhile hearing from time to time sinister stories of how Elites fell apart at the seams after a few months' driving on rough roads, and so on.

Then a few months ago I saw it—a picture of bits of Elite spread out all over the ground. The vehicle was in future to be made available in component form, thereby discarding its eligibility for purchase tax for anybody who had £1,300 to spare and the urge to have a crack at the mother of all do-it-yourself kits.

It needed the Motor Show to confirm the intention. Many minutes were spent in a close scrutiny of both the bits and the completed car which could be found on the Lotus stand. Eventually the die was cast, and on a wet and dismal morning in early November the Lotus was delivered to the house of a co-operative brother-in-law who had considerably better facilities than the writer for undertaking the task of assembly. The manner of the Elite's arrival was amusing, for the Lotus van which brought it contained only the mechanical components, while the body shell was towed on a trailer behind.

Heeding the warning of the makers, three strong helpers had been laid on to help unload the pieces and the body shell. It is an essential precaution if the body is to be handled properly and not strained by lifting at the wrong places. As far as obtaining helpers is concerned, one of the delightful things about a job of this sort is that assistants and advisers are extremely easy to come by. The parts were all checked off against a list provided and the whole business of unloading was accomplished in about fifteen minutes.

Everything for basic construction had arrived safely, although the heater, which had been ordered as the only item of optional extra equipment, had not been included. A brief check with Lotuses indicated that there was a temporary shortage of heaters, but that the equipment could be added at a later date. With all the components stowed around the garage it looked as if it would only be a few minutes work to string the vehicle together. The body shell, for instance, had arrived complete with interior trim, instruments in place, wiring harness, fuel and brake lines attached. Even the seats were inside although they were not mounted to the floor. Of the mechanical parts there were four sub-assemblies of the suspension parts, the bare engine and gearbox with a box containing auxiliaries, five wheels, an exhaust system and so forth. Appearances in this case proved slightly deceptive in that the suspension sub-assemblies require a certain degree of dismantling before they can be put into position. Similarly, although the fuel tank is bolted in position in the boot it is necessary to remove this in order to attach the differential which, with the independent rear suspension, is bolted direct to the glass fibre body shell.

Initial discussions took the form of deciding what should

Temporary resting-place on the garage floor for the body shell before the building work began

Just a heap of bits in the back of a van. Creating order from apparent chaos was not as difficult as it might seem

The long-awaited arrival of the kit of parts. Getting the body off the trailer without damaging it required four strong men and a deal of thought

All the initial work was done with the car up on packing-cases. With suspension, wheels and differential in place the shell was lowered before fitting the engine and gearbox

Slave at work. Getting the spacing washers into place in the rear suspension assembly required considerable patience

be done first. One school of thought contended that the engine should be put in place first, while the other favoured mounting the suspension so that the vehicle would be manoeuvrable for installing the engine. In the end the latter course was adopted although it was found with some dismay that advice given in the comprehensive workshop manual indicated that the first course was the more correct. By the time this was discovered things had advanced to such a stage that it was well worth a try to get the engine into place without removing all the suspension. The workshop manual, incidentally, gave some invaluable assistance through drawings and the accident repair section when I was in doubt.

The provision of the suspension components in sub-assembly form quickly proved its value, for it was possible to inspect the exact method of assembly before undoing the various parts to fit them to the car. Since it had been

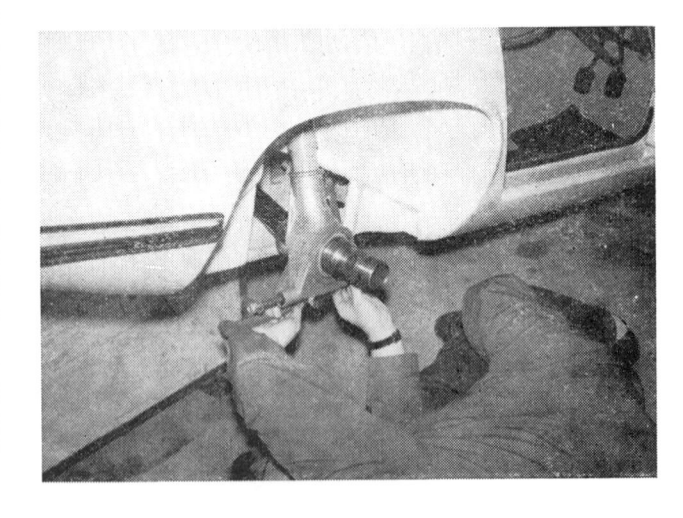

The wheels were supplied shod with Pirelli 'Cintura' tyres, but had not been balanced, a job which will have to be done before much high-speed running is indulged in

When mounting the differential (seen here from below) the brake discs were removed and refitted with the centre case in position. The aluminium heat-shields on the body had to be adjusted to avoid fouling the discs

decided to put the suspension on first the willing helpers and myself managed to lift the complete body shell onto a couple of packing cases which allowed a generous amount of room for working underneath the car. Subsequent experience showed this to be of considerable help. On the underneath of the car there are marks painted to show the points at which the structure is capable of carrying its own weight. By attaching the suspension before installing the engine and by the use of packing cases we were able to spread the weight over a larger area and at the same time keep that weight to a minimum. In fact, our boxes were not put under the recommended marked places because it was found easier to work slightly differently.

No difficulty was encountered in the mounting of the rear suspension until the wishbone had to be attached to the base of the patent Chapman suspension strut. A certain amount of patience and jiggling is required to pass the bolt through the end of the wishbone arms, picking up the intervening washers on the way. It is also necessary, incidentally, to ensure that the wishbone has been put into position before the suspension unit is attached. The differential was then bolted into position, and this operation is one of the most difficult of the entire assembly because there is insufficient space to work on the bolts. Two people are essential for the operation, one underneath the car to put the differential into position and fit the nuts on as the bolts are pushed through from the luggage boot side. The two upper bolts could be tightened without very much trouble, but for the others the poor underneath-man had to twist his arm through the small gap between the differential and the disc brake calliper to attempt to get a spanner in a position to hold the nut. Eventually this was achieved but not without some loss of patience. The drive shafts were then bolted through the disc brakes, again a fiddly job, and it was found that it was impossible to lock one of the nuts because the bolt had not been diametrically drilled and the split pin could not therefore pick up the groove on the other side of the nut.

The front suspension posed fewer problems, although we found it necessary to remember to get the shock

Fitting the engine/gearbox assembly. A chain block was found to be a great help at this stage but care was needed to avoid damaging the edges of the bonnet opening. The top illustration shows the engine suspended at approximately the right angle while the protective tape is removed from the gearbox output shaft. The bottom picture shows a half-way stage and in the centre the engine is home and dry

Nearing completion. The radiator is in, carburettors and manifold have been fitted and the pipes and cables connected up

With the seats in and interior trim in place the car is almost ready for the road. During our handling of the body shell one door was slightly strained and became difficult to shut—door hinges are a slightly tender spot of the Elite

absorber inside the wishbone before starting to do things up. It is important to remember that final tightening should not take place at this stage either for the inboard end of the rear wishbone or for the front wishbone because it is necessary to have the car in a loaded position for this operation.

With the differential in place it was possible to put in the fuel tank once more, coupling up the fuel line and the fuel gauge wires. Here we came across one of the very few examples of poor fitting where the neck of the fuel tank did not line up with the filler cap on the outside of the bodywork. A piece of flexible rubber piping had thoughtfully been provided, but even so the discrepancy was such that it took a deal of fairly strong-muscled manipulative work to achieve a connection between the two. By this time a couple of afternoons and evenings had been spent working on the car to a total of some twenty man-hours and the time had arrived for the major task of installing the engine. This task was approached with some trepidation in view of the sequence in which things had been done. The likely problem appeared to be the fitting of the prop shaft into the rear of the gearbox, for the prop shaft tunnel is entirely enclosed except for one small opening opposite the end of the gearbox when the engine is properly mounted. Nevertheless a block and tackle was borrowed from a willing local farmer, and this was attached to a girder provided for just such a purpose at the entrance to the garage.

With the suspension in place it was possible to afix the wheels (after applying a little graphite grease to the splines), move the packing cases and let the Lotus stand on its own. We were then able to manoeuvre the car forwards and backwards for the easier installation of the engine. In this respect our sequence of operations was undoubtedly superior to the alternative method of installing the engine first. Rope slings were made for the engine and put in the position recommended in the workshop manual for installing and removing the engine. The first time our calculations were not quite correct, and when the engine was lifted on the block and tackle the degree of tilt was far more than the 30 to 35 degrees recommended to enable the engine to be lowered through the rather small bonnet opening. A quick adjustment left us with something in the region of a 25 degree tilt and we decided to go ahead in view of the fact that the car was on a slight slope out of the garage. Once the car is in position under the engine it is a question of lowering gently and then moving the car forward an inch or two at a time. At least three people are required for this operation, one to lower the engine, one to guide it at the bonnet end, and a third to feed the gearbox through the mounting points and onto the propshaft. Care has to be taken that when the engine swings it does not touch the sides of the bonnet opening, because even a slight bang suffices to chip off the paint. Inch by inch the engine was lowered and with surprisingly little trouble the propshaft was inserted into the gearbox. Before the weight of the engine was released the mounting bolts were slid into position and the nuts loosely run down. The actual job of installing the engine had taken some fifteen minutes although, of course, the preparation of the slings and block and tackle had taken considerably longer.

Now began the numerous fiddling jobs of connecting up all the auxiliaries, pipes and what have you. The inlet manifold bolts presented something of a problem to tighten up but this was eventually achieved, one of the main drawbacks being that it is not possible to reach anything on the engine from underneath for the engine compartment is completely sealed in except for a hole through which

After fitting the gear lever the hole in the tunnel trim had to be opened out somewhat to enable all the gears to be engaged

Brake and clutch reservoirs and engine oil filler are grouped conveniently, and from somewhere among the confusion of pipes and cables the throttle control emerges. Arranging the proper operation of this control took some time

the engine sump protrudes. Two more examples of inaccurate fitting were discovered when it was found that only three of the radiator holding bolts could be affixed since the fourth hole was not in the correct position; and part of the exhaust pipe had to be sawn off. The throttle control was particularly tortuous and it was a long time before this could be made to operate satisfactorily. The brakes and clutch were connected up and bled, the various oils inserted, fuel was put in, and the grease gun was taken around. Here occurred our first serious mishap when during the process of greasing an inboard drive-shaft universal the nipple broke off, and it proved impossible to rescue the threaded end. The interior trim was now restored, much of it having had to be removed for the installation of the propshaft and gearbox; the speedometer was connected up, all the other instruments having their connection made at the external end; and finally, after a precautionary trickle charge, the battery was put into position. There is one curious feature of the interior trim, which was re-affixed at this stage. The hole provided for the gear lever is not sufficiently large to allow the gears to be selected, and this must be enlarged. A key-hole saw proved the ideal tool for the job and the work was accomplished in a few minutes.

When the two battery cables were attached, to our chagrin all the lights on the car came on, including the indicators—both sides at once! A quick investigation showed that the fuel gauge was registering full when there was only a gallon of fuel in the tank, and various other anomalies. Although the workshop manual indicated that the auxiliary lead on the battery should be attached to the positive terminal, it was found that when this was put on the negative side the lights and most of the instruments worked correctly. The ammeter persistently showed a charge, however, when it should have been at rest, and the fuel gauge continued to behave rather erratically.

Now came the great moment when an attempt was made to start the car. Unlike most of these ventures which are crowned with scintillating success when the engine fires at the first touch of the button, the Elite's Climax motor stubbornly refused to start under any circumstances. Fuel and electrics both appeared to be alright although the spark was rather weak. Eventually the fault was traced to the distributor, inside which the lead to the condenser had become disconnected. This probably occurred when the low tension wire was connected up and the bolt shifted, thereby deranging the condenser. When this had been rectified, sure enough the engine started, although it had far from the customary Climax crackle, and it was obvious that the carburation was sadly awry. Eventually this too was adjusted and there was a complete Lotus Elite, resplendent in white paintwork and tan upholstery waiting to be registered and driven away. The replacing of the bonnet lid, incidentally, is a fiddling business because in order to get the best possible fit the hinges and the mounting flanges on the bonnet have been slotted. It is a lengthy business to do a two plane adjustment on each side so that the bonnet lid is squarely in the bonnet hole and flush with the sides as well as not too stiff or too slack in the locking mechanism.

The formalities of registration were not familiar, but the application for a licence was obtained from the post office and sent, together with the bills as instructed thereon, and in a couple of days the licence arrived back. The bills have not been returned at the time of writing so presumably there is some liaison between the licensing authorities and the purchase tax officer regarding the clearance on the tax question.

But the car was now licensed and arrangements were made to have it returned to the Lotus works where a free inspection is given to ensure that the car is fully roadworthy. The short drive entailed in getting the car to Cheshunt was a delightful experience. Even if a car provided in this form is not, as the ads imply, as easy as a jig-saw puzzle, it is not a difficult nor a tedious business. Yet the very fact of being involved in its birth brings out the father in a man. It should be mentioned in this context that approximately 45 man-hours were spent on the basic assembly, discounting time on fiddling and adjusting. This is nearly twice as long as claimed in the aforementioned advertisements—but then neither myself nor my helpers could remember ever having done a jig-saw in our youth.

CASE HISTORY

Three months with the Elite

Bill Dale-Otis

On the road and fully run-in—with symbolic derestriction signs in the background!

IN my story of the building of my Lotus Elite it was not possible to give more than a very sketchy impression of what the car was like on the road, since it was necessary to publish the article in the Racing Car Show issue and the vehicle was only first licensed on December 1, 1961. Some ten weeks have passed since that day. Now I put pen to paper again in order to cover the first 3,500 miles of the car's life.

A fair amount of amusement was caused when it came to making up the number plates for the Elite. The rear one is, of course, perfectly straightforward, but at the front there are two alternatives, neither strictly conventional. The plastic type of plate can be stuck on the bonnet top, or the letters can be mounted directly onto the fine mesh grille in the radiator orifice. Neither method is strictly speaking legal, but of the two I felt that the latter was not only smarter but more nearly in accordance with the letter of the law than the plastic type. The grille was accordingly taken along to a local number plate maker-upper and, after some discussion, it was decided that a white polythene letter would look smartest in conjunction with the white body. This letter has an extrusion on it which could be pushed through the mesh of the grille and secured by a gripping washer applied by a special tool. Needless to say, the extrusions on the letter did not always coincide with the holes in the mesh, but the flexibility of the plastic allowed sufficient latitude to get them all in place. The grille was carried home in triumph but to my great dismay, on coming to put it into position it was found that the letters had been put on the wrong side, there being no way of telling the difference except that the fixing tabs on the grille should be facing backwards. All the letters and numbers had to be dismounted and fixed on the other side without the help of the quick-action tool. Moral: it pays to be careful.

As mentioned in the earlier article, the car was now taken back to Lotuses for a post-construction check-over. The main doubt in my mind lay in the rather excessive noise which appeared to be coming from the differential. It seemed likely that this might have been bolted too securely to the glass fibre body, thereby reducing the damping effect of the rubber bushes. To slacken off these nuts would involve removing the fuel tank from the boot. But it turned out that the rather excessive noise in bottom and reverse gears was fairly normal, and the car also passed the other tests with flying colours. A grease nipple had been broken off one of the inboard universal joints during greasing, a comparatively easy thing to do because the clearance is very marginal and the drive shaft has to be rotated to exectly the correct position to get a grease gun onto the nipple at all. Having done this a hasty withdrawal of the gun can cause the nipple to snap off. The end could not be extracted, so it was necessary to fit a new UJ.

The car was returned once more to Lotuses for the 500-mile service and in the hope of having the heater and demister unit fitted The latter was not available and the servicing almost ended in disaster when the car was released late one evening from the Lotus service department with no oil in the sump. It was driven for a couple of miles before the fluctuating oil pressure indicated that something was seriously amiss, and a hasty return was made to the Lotus factory. Matters were promptly put right and there was no immediate indication that any permanent damage had been done to the engine. Time alone will show whether in fact this is the case, short of a complete strip down and examination.

It was now mid-December, and although the car had been on the road for a fortnight the bills which had been sent to the licensing authorities had not been returned. Then, one afternoon when the car was not at home, a gentleman from the Customs and Excise department turned up at the front door and requested to inspect the Elite in order to ensure that it was indeed a home constructed vehicle. My good lady was unable to oblige, but a date was made for a week thence, and the gentleman duly turned up once more, looked at the car, looked under the bonnet to check the engine and chassis numbers and departed. A few days later the bills were returned and the log book sent with the cryptic note on in red letters "built up vehicle".

By the time Christmas arrived the car was very nearly completely run in and the delights of a superb little vehicle were making themselves felt. The corks were really blowing out of the engine and, as it became possible to take the revs higher, the real power section of the rev range could be experienced. It is an interesting thing, incidentally, that for running-in Coventry Climax recommend frequent steady loading of the engine in the intermediate rev range rather than over-sparing use of the throttle. Early experience with the Climax engine in production cars apparently produced abnormal fuel consumption figures. Since no such problem had ever occurred with racing engines it was concluded that fiercer running-in was beneficial to the engine in that bore irregularities were removed rather than being surface polished.

This procedure was followed carefully but the oil consumption did quite suddenly take an upward turn, and it was discovered that oil was leaking from the cam box cover in quite considerable quantities. Although a new gasket was fitted the fault still remained, although in a more moderate degree. Again, racing practice may come to my aid in that the cars when raced use a cam box breather to relieve pressure. About the same time the joint between the exhaust manifold and the down pipe blew. Fitting a gasket in place here was a tricky job, because the carburettors prevent easy access to the nuts. It was achieved eventually, although whether entirely satisfactorily or not is doubtful since a few hundred miles later the gasket blew again. Once more it was replaced and so far the car has covered another few hundred miles without giving further trouble. I recall an exactly similar experience with one of the early Austin-Healeys which was very prone to blowing this particular gasket. It may well be that the high performance four-cylinder engine as a species tends to have this particular fault.

By the end of the Christmas holiday the Elite was fully run in. Now it was possible to see the rev counter needle bounding joyously up towards the 6,000 mark, and to feel the fierce surge of power from the modest 1200 ccs over the whole latter part of the rev range. But with the excitement and the performance came an acute awareness of the noisiness of the car. The Climax power unit is not a quite one and it seems that the glass fibre body shell of the Elite is a particularly good sound conductor, if not amplifier. High on the list of priorities is a visit to Interior Silent Travel to see if improvements can be effected in this sphere. Bad weather over the Christmas and New Year periods allowed the Lotus to be extended on slippery and, at times, icy or snowy roads. It was proved to me beyond a shadow of doubt that the suspension system is absolutely right. The car was tremendously manageable, and one occassion which sticks in my mind is worth recording. I was following a couple of cars along a winding road at night just after Christmas, and I could not understand why both were proceeding so gingerly. The road was lit by street lamps but not particularly well. Nevertheless I was not using my headlamps because of the incovenience I would thereby cause to the cars in front. Eventually a sufficiently long straight appeared for me to slip into third gear and accelerate past the two slower cars. As I did so I switched on my headlamps, and when I did so I realized the reason for the slow progress. Back from the road surface twinkled the unmistakable sparkle of ice, I was by now committed to the overtaking manoeuvre and carried on, but with same caution. Once past the other cars a little experiment showed that the road was indeed icy, but the Elite was taking things in its stride in the most amazing manner, and it required really harsh movements of the steering wheel or throttle opening to induce the front or rear wheels to lose their adhesion Part of the credit for this will undoubtedly go to the Pirelli Cintura tyres that are on the car, with their ingeniously-devised braced tread construction.

Although it is not an easy car to get in and out of, once inside the Elite everything feels absolutely right. The visibility forward is superb, the seats grip in all the right places and the dashboard layout and the controls are all arranged in a businesslike fashion. Rearward vision is limited by the top of the rear window, and the lack of an opening side window is a trifle irksome at first. For those unfamiliar with an Elite it is necessary to point out that the car has no winding windows, merely a removable pane. One of the disadvantages I have discovered about this removable pane is that the car must be stationary and the door open in order to extract it. Thus, if you find the ventilation is inadequate as you are driving along, it is not possible just to pull the window out. But whatever way you look at it, the basic shape of the Elite removes any practical possibility of having winding windows, and the result is a very reasonable compromise.

The very compact dimensions of the car make it a very selfish little vehicle; it is pretty tight going for two people and ordinary hand baggage! Another thing that I found difficult to get used to is the combined headlight flasher and horn which are incorporated on a stalk type switch to the right of the facia panel. Downward pressure on the stalk produces the horn and upward a flash of the headlamps. I have found it rather easy to get horn when I wanted headlights and vice versa, or even both apparently at the same time. Another disadvantage is that it is much easier accidentally to bang a stalk than the conventional button, thereby drawing disapproving glares from fellow motorists and pedestrians alike.

All in all the Lotus has given satisfactory service and a great deal of pleasurable driving in the first two and a half months of ownership. It is, of course, too early yet to pass any observations on the car's robustness and general quality of construction. The only defect I have noticed so far, and this may be due to inaccurate assembly, is that the bonnet seems to have bowed very slightly so that there is a gap in the centre of the trailing edge which was not there when the bonnet was first put back in position. I am keeping a good eye open for cracks in paintwork and other signs of bodywork deterioration which have been the subject of complaints from some readers of *Sporting Motorist*. Meanwhile, if I can hark back to the words of the company's advertisements, I am enjoying being one of the elite.

The Lotus that Came in a Crate

By Dennis May

A British physician buys a Lotus Elite in CKD kit version

Neighbors helped unload the fiberglass body shell of the doctor's Lotus Elite. It was surprisingly complete, with lights, bumpers and instruments installed.

It seemed more like a three-dimensional jigsaw puzzle than an engineering project when the components were laid out.

Dr. Kenneth MacDonald, of Hook, Surrey, England, loves sports cars, particularly the new Lotus Elite. But, like many British enthusiasts, he cannot afford both the car and the tax Her Majesty's government puts on it. An Elite costs £ 2,006, of which £ 631 is tax.

Fortunately for the good doctor and for many other drivers in England, Colin Chapman, Lotus's boss, relieved them of their dilemma by putting out an Elite in kit form, thereby making it possible to sell the car in assemble-it-yourself condition for just £ 1,299 tax-free.

MacDonald is 48, married, and has two daughters. The younger of them, 17-year-old Jackie, is a technical-college student; she happily agreed to contract a convenient ailment when the kit's delivery drew nigh, and time her convalescence to coincide with the assembly job. What she, her father and a friend did can be seen on these and the following two pages.

At the planning stage he had two problems—working space and man power. You could swing a cat *or* accommodate a car in the MacDonald home garage, but not both at once. As to labor, Ken had

56

freed his own hands by engaging a colleague to attend to Hook's health for four or five days. But it was obvious that Jackie couldn't help with the heavy work.

Tommy Atkins, who fortunately has his garage nearby, and is well known for his kindness and helpfulness, came to the rescue. Bruce McLaren was racing Atkins's Cooper-Climax in the Antipodean Temporada, and High Efficiency Motors's small racing shop, just across the street from the doctor's place, had space to spare for the Elite. MacDonald could also have the part-time services of H.E.M.'s apprentice racing mechanic, 17-year-old Peter Rambuat.

Delivery of the kit went according to plan and involved no transit damage. Off-loading, with the help of local well-wishers, took about ten minutes. The shop was big enough for two cars to be worked on simultaneously, and storage of the Elite parts offered no problem.

PKD (partly knocked down) would be an apter abbreviation than CKD (completely knocked down) for the state of untogetherness in which this Lotus set reaches the customer. The fiberglass hull comes completely color-finished, glazed, trimmed, and with its doors, electrical components, wiring harness, and steering gear fitted.

The disintegral hardware therefore consists of an engine-gearbox unit, front-suspension sub-assemblies, rear-suspension tackle, a differential unit (incorporating the rear brake discs), driveshaft, fuel tank, five road wheels, an exhaust system, hood and trunk lids, two seats, a box of assorted fastenings, fittings and other bits.

Ken had ordered the rare but optional ZF gearbox, which has close ratios and synchro on all four gears rather than the upper three in the standard MG box. He'd also arranged to have his Climax engine worked over by Jack Brabham's specialists before installation.

Rigging the rear suspension presented few snags, all in all, apart from trying to do the job with the car so low to the garage floor. True, they prematurely tightened and wire-locked the bolts retaining the cups for the radius-arm ball ends, but this was due to momentary inattention to the book.

The positioning and attachment of the upper ends of the spring/shock units being Phase One of the suspension-rigging operation (these units are located within the protuberances, officially termed metacones, that meet your gaze as you look in through an Elite's back window), gravity at no time frustrates your labors in this area. Any juggling with the half-shaft and radius arm that may be necessary is thus much facilitated.

Work outlined so far, plus many minor chores that lack of space compels us to leave to the imagination, took up the first 10½-hour day.

Straightening out cricks in their necks that were threatening to become permanent, caused by long-time crouching at sub-Elite level, Ken and Peter relievedly resumed their natural stature and called for a recitation of the gas-tank-fitting formula: No theory-*vs*.-practice clashes here, thanks to the book's detailed and succinct instructions for avoiding the one potential peril, kinking the flexible fuel line, the tank went in easily.

Connecting up the front brake hoses and bleeding the whole system came next—the lines to the

Daughter Jackie reads instructions for the propeller shaft to her father while he cleans the universal-joint flange.

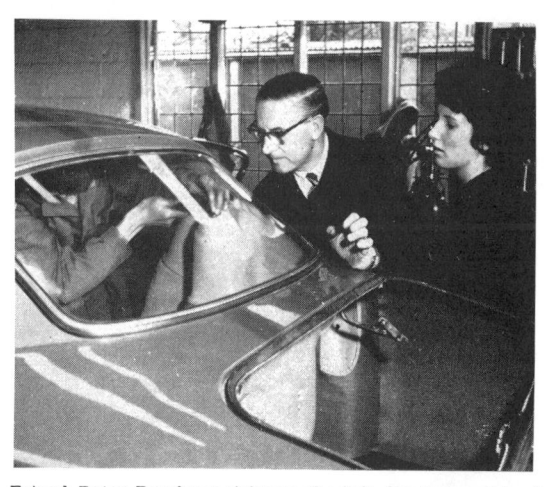

Friend Peter Rambuat tightens the bolt for the combined spring and shock-absorber unit from the inside of the car.

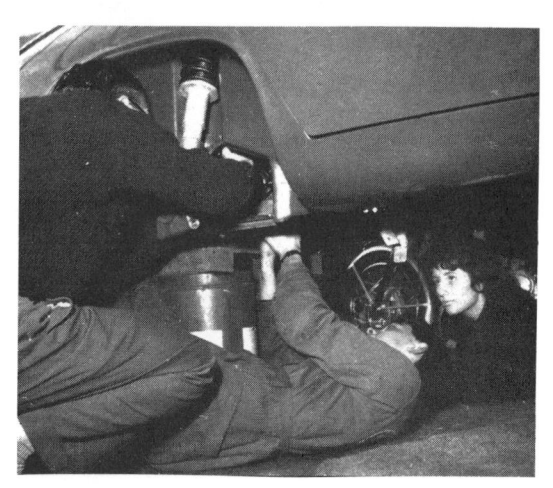

The installation of the final drive and the rear brakes and the mating up of the half-shafts proved to be a simple job.

Almost without blood, toil, tears and sweat, the doctor and his helpers completed the assembly of a Lotus Elite components kit

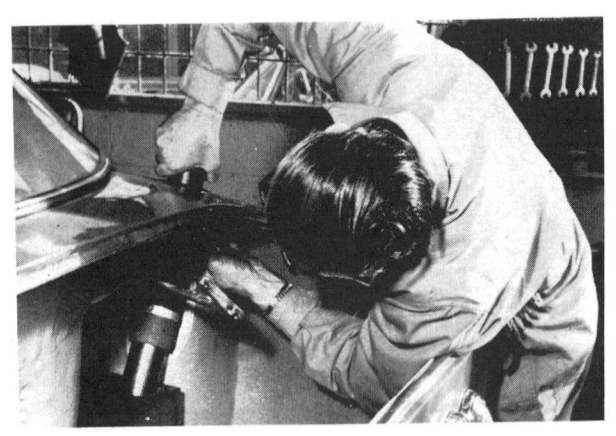

An early step in the assembly process was to install the gas tank. Here the filler neck extension is being fitted.

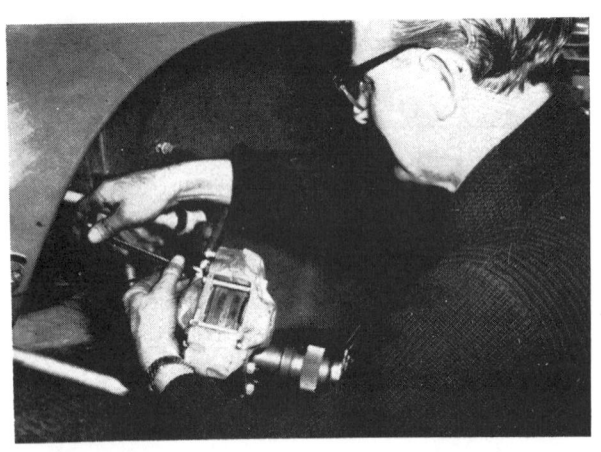

Dr. Ken MacDonald completes the brake system by attaching the brake fluid line to the 11-inch front disc brake unit.

inboard rear brakes having been coupled previously. That completed, the road wheels were fitted and the car was ready for lowering onto its tires.

At last the gearbox showed up, preceded by the redelivery from Jack Brabham Motors of the worked-over engine, and next day after breakfast the FWE Climax was duly mated to the ZF.

With the engine-gearbox unit finally homed, the nuts tightened at its three mounting points, and connections made at either end of the driveshaft, the rest was easy. The radiator fell into place on the front sub-frame, its attachment bolts and washers were exactly where we remembered stashing them, and the captive nuts on the radiator shell aligned perfectly with the holes in the mounting lugs. The cooling system jig-sawed together just as the book said. The same mood of sweet reasonableness infected the electrics, the exhaust downpipe, throttle cable, choke, tach and speedometer drives, oil line.

Throughout the job the team had been impressed by the fact that no fiberglass surgery was necessary; with the exception of the aperture to the gearbox filler plug all holes and breachings in the hull structure, including its various internal walls, had proved accurately placed and sized. Otherwise, there is just one point where you must resort to cutting—the top of the transmission hump. To quote the Lotus literature: "Before the gear lever and grommet are fitted, the 1½-inch-diameter hole should be opened out to three-inch diameter in such a way that the lever will engage all gears . . . and still leave approximately ¼-inch clearance with the edge of the trim."

About the only thing that marred the first rapture of owning and driving a self-assembled Elite was a high level of resonant noise within the body. This is, of course, a familiar acoustic phenomenon with i.r.s. cars with monocoque bodies, caused by magnification and direct communication of transmission hubbub to the hull or body. Ken pondered the matter, then tried underlaying the trim on the driveshaft tunnel and spare-wheel platform with thick felt. This abated the din by at least half, at negligible cost.

Owner-assemblers of Elites are encouraged, if their lines of communication are short enough, to report with their cars to the Lotus service department for a free check and road test after the first 500 miles. Ken took advantage of this and was impressed by the thorough vetting.

Three faults came to light: one, the plastic fuel line, all our care notwithstanding, had been kinked and needed replacement; two, a rear brake line had been misdraped and required slight re-routing; three, last but far from least, the front anti-roll torsion bar hadn't been assembled upside-down by Lotus, as we had thought because of difficulty in attaching it, and therefore had to be re-inverted. It remains a mystery, looking back, how the related suspension components could have been made to jig-saw together without upside-downing the bar, but, mystery or no mystery, we stood corrected. The rectification Lotus carried out here eliminated 240WPL's sole steering idiosyncracy, a slight wandering tendency at speed.

Ken calculated he'd put in around 36 hours' work, spread over the two whole days and varying fractions of three more days.

His last word: "What I'd really like to do is build two. Knowing what we know now, we'd get the second one together in not much more than half the time this one took." It could be said that at the rate of £700-odd saved on every Elite you build yourself, you can't afford less than two. **C/D**

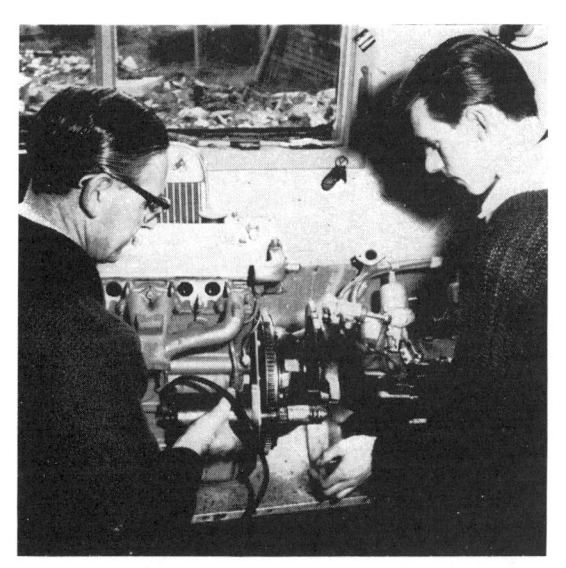

It was a big moment when the doctor and Peter Rambuat got the ZF gearbox mated to the Brabham-tuned Climax.

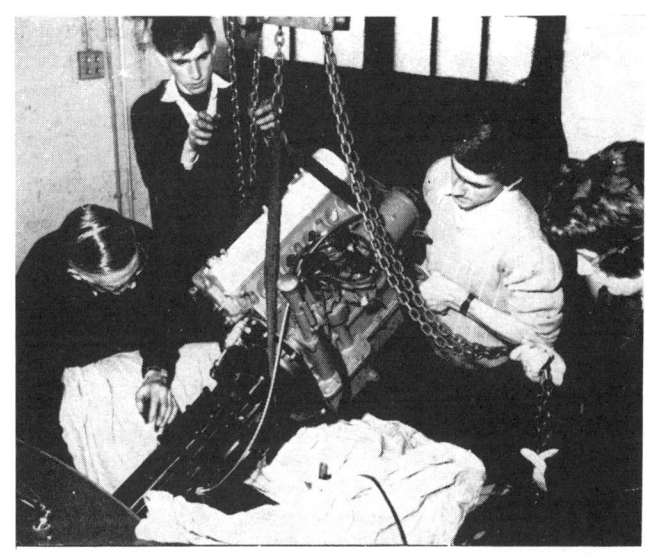

Dropping the engine and gearbox into the car was, contrary to the factory's instructions, one of the last steps of assembly.

After a successful trial run, the car was posed with attendant obstetricians Jackie MacDonald, Peter Rambuat and owner.

REASSESSMENT
THE LOTUS ELITE

BY DAVID PHIPPS

ABOVE: Colin Chapman with the prototype Elite first shown at the 1957 Motor Show.

I FIRST drove a Lotus Elite early in 1958, some six months after the model's rather premature announcement at the 1957 London Motor Show. At Earls Court, even though incomplete and untested, it was generally considered the Star of the Show. Driving it for the first time I was able to overlook such inconveniences as excessive noise and ill-fitting doors; the Elite more than made up for such things by its driving position and its handling, its performance and its brakes. Within a minute after taking over the car I was doing 100 m.p.h. in complete comfort, free from any of the tenseness which is normally associated with the first few miles in a strange vehicle. Cornering was accomplished by inclination of the wrist, braking by resting the right foot on the central pedal. The engine roared, the final drive gears whined and there was a tremendous resonance in the roof each time the throttle was eased, but who cared—here was a racing car disguised for use on the road, and very beautifully disguised too.

In March, 1959, I contrived to be going to Geneva at about the time an Elite had to be delivered there. It was built the day before I planned to leave and given a quick test "round the block". Everything seemed to be in order except that the petrol gauge didn't work, and I was told to consider the car run-in; all Coventry Climax engines are run on the bench for several hours before being delivered.

The next morning I drove to the Silver City air terminal at Ferryfield, the Elite skimming over the last few miles of Romney Marsh at around 100 m.p.h. After a break of half an hour, during which it was whisked across the Channel

LEFT: The currently popular "Do-It-Yourself" component Elite which was introduced in 1961.

by Bristol Freighter, passed through two sets of Customs (neither of which managed to find the concealed bonnet-opening catch) it carried on at the same pace on the deserted roads of Northern France, even on the cobbled, steeply cambered section near Arras, where it ran straight and true at 110 m.p.h. at the expense of considerable noise from its Michelin "X" tyres. Soon after this, however, I was to regret the inoperative fuel gauge as the engine cut—almost without warning—and the car coasted to rest; the top of the carburetter float chamber had vibrated loose and a tankful of best 100 octane petrol had been pumped out, odourlessly, on to the road. Fortunately a motoring St. Bernard, in

RACING ELITES: Graham Warner's "LOV 1" and Les Leston's "DAD 10" enlivened many G.T. races last season when they were engaged in fierce duels.

the shape of a Simca owner with a jerrycan, helped me out of my predicament, and with the engine pinking merrily (it was French "regular" petrol, which is about on a par with paraffin in octane rating) I made my way to the next filling station. A tankful of mixture-grade "Super" fuel reduced the clamour to a bearable level, and less than three hours after leaving Le Touquet the Elite rolled into Rheims—an average speed of over 60 m.p.h. despite the stop.

The rest of the journey was remarkably uneventful, considering the combination of strange roads, a new car and a very high cruising speed. There were no alarms or excursions, no panic stops, and even when the Elite suddenly came upon a corner at the end of a long straight it always went round it quite impeccably. With nightfall, hilly roads and finally snow, the average speed slackened somewhat, but even when the road was completely snow-covered the car refused to misbehave. On arrival in Geneva the only major fault I had to report was uneven part-throttle running, due to the use of ultra-soft engine mounting rubbers. For most of the trip I had been unaware of this as the throttle had been wide open; even so, fuel consumption after my enforced stop worked out at 35 m.p.g., a real tribute to aerodynamics and light weight.

During the following year I drove several more Elites, all of which had been subjected to intensive development —with varying degrees of success—in an attempt to lower the noise level. Part-throttle running was much improved and suspension noise was reduced, without adverse effects on the wonderful handling. The exhaust resonance remained, however, and the engines were subject to vibration periods—particularly at just under 5,000 r.p.m., which is the car's normal cruising speed on most British roads. In addition the gearshift was extremely stiff, and there was no heater or demister. I reluctantly came to the conclusion that it would be impossible to make a "refined" motor car of the Elite without seriously impairing its performance and handling. It was too dear, anyway.

After this I didn't drive an Elite for over a year, and then—soon after the "Do-It-Yourself" version was announced —I was able to borrow one for a whole week. I was glad to find that all the old responsiveness remained—in fact the roadholding was if anything even better than before, thanks to the fitting of Pirelli tyres. These are heavier than the Firestones previously fitted as standard but have excellent gripping qualities on both dry and wet surfaces. I was able to compare the two when the right rear Pirelli picked up a nail in its sidewall during some high-speed cornering. The Firestone-shod spare, although very good by normal standards, could be set sliding far more easily than the Pirelli, particularly in the wet, which made extra concentration necessary on left-hand bends. Another illuminating aspect of this enforced wheel-change was that the spindly handle supplied with the scissors-type jack twisted up like a piece of wire; however, the Elite is so light that it was quite easy for me to hold up the side of the car while my wife slipped one wheel off and the other on.

This apart, the Elite behaved itself extremely well. It started instantly whether cold or hot and warmed up very quickly. The doors opened and shut easily, there were no draughts and no leaks. Perhaps the most surprising feature, however, was that the car was so much quieter than the earlier examples. Gone was the roar of the engine, the thumping of the suspension and the resonance of the exhaust on the over-run. This Elite was by no means silent, but the general level of noise was no higher than that of most sports saloons, and lower than that of all soft-top production sports cars.

Another pleasant surprise was the lightness of the gearchange, which has thus been brought into line with all the other controls. The steering seemed almost incredibly light—perhaps too much so, for it lacked self-centring—and the well-spaced-out pedals all functioned admirably with very moderate pressure. The brakes at first felt rather dead, but proved capable of stopping the car time and again in most reassuring fashion. The various dashboard switches also worked well, particularly the single knob which pulls out to operate side- and head-lamps and twists to vary the intensity of the instrument lighting or switch on the roof-mounted interior light. And the headlamp flasher/horn blower stalk, ideally placed just below the steering wheel at about two o'clock, makes it possible to give audible and/or visual warning of approach without removing either hand from the wheel rim. The heater/demister is very effective, but the pull-out heat control is liable to foul the driver's knee.

The one major snag which remains— to judge from the behaviour of the test car, which should be a better-than-average example—is unbalance of the crankshaft, flywheel, clutch assembly, which leads to considerable vibration at high speed. That this is not an unavoidable adjunct of motoring behind a Coventry Climax engine is shown by the fact that balanced units, as fitted by Jack Brabham Motors in Triumph Heralds and M.G. Midgets, are extremely smooth throughout the full r.p.m. range. For anyone prepared to assemble the car really carefully, and to pay a little extra for balancing the engine and clutch, the Do-It-Yourself Elite is extremely good value for money. There is certainly not a production sports car to compare with it.

LOTUS ELITE bodies coming off the production line at the Plastics Department of Bristol Aircraft.

61

Equally at home on road or track, part of the Elite's road test was conducted at Brands Hatch circuit, in Kent.

LOTUS ELITE (Special Equipment)

ALTHOUGH there has been little apparent change in the Lotus Elite since its introduction in 1957 with a form of construction which was unique, it remains one of the world's most advanced small Grand Tourers. A stressed glass fibre shell provides the basis of the car, with independent suspension of all four wheels. The Special Equipment Elite has a more highly tuned version of the 1,216 c.c. Coventry Climax 4-cylinder engine and an all-synchromesh ZF gearbox. Although noise has still to be defeated, the Lotus Elite is a superlative car by reason of excellent roadholding and steering, an outstandingly level and well-damped ride, good, although very heavy brakes and a conspicuously rapid performance. Its clean lines justify both aesthetically and aerodynamically the good reception they were given when the car made its first appearance, and the interior is as smart and comfortable as any current production. The price in completed form is not low, but when taken as £1,451 in kit form, which is how most Elites are sold in this country, it is certainly not excessive. The high noise level is a tiresome drawback of a car which is otherwise a delight to drive.

Rigid and well-balanced

THE only metal in the hull is a steel tubular hoop in the roof. The shell is extremely stiff and there is little in the car's behaviour to suggest that the structure is unusual. The front suspension is by wishbones and coil springs and the back is by coil-spring-and-damper struts with trailing radius arms. The construction of the Elite in this manner has the virtue of light weight as well as simplifying assembly by the amateur. The single overhead camshaft Climax engine in the special equipment model we tested has two SU carburetters and produces 85 b.h.p., indicating that it is still working well within its known capabilities. Disc brakes all round, those at the back being mounted inboard at the differential, complete a specification which is as ingenious as might be expected coming from the drawing board of one of the most talented and successful manufacturers of sports and racing cars in the world.

The performance of the Elite comes from an engine which, while it is smooth and powerful, only gives its best when it is

In Brief

Price (in kit form as tested) £1,451.	
Capacity	1,216 c.c.
Unladen kerb weight	13½ cwt.
Acceleration:	
20-40 m.p.h. in top gear	9.5 sec.
0-50 m.p.h. through gears	8.2 sec.
Maximum top gear gradient	1 in 9.2
Maximum speed	117.5 m.p.h.
Overall fuel consumption	27.5 m.p.g.
Touring fuel consumption	41 m.p.g.
Gearing: 16.75 m.p.h. in top gear at 1,000 r.p.m.	

A wood-rimmed steering wheel and clever use of heavy, black p.v.c. make the interior comfortable and stylish. The lever projecting from the facia conveniently near the driver's right hand works horn and headlamp flashers. Small buttons on the toe-board behind the pedals are for screen washers and headlamp dipper.

Removable Perspex windows in the Elite stow in pockets in the seat backs. They are easy to take out and replace, even from inside the car.

LOTUS ELITE

revved freely. Power below about 2,000 r.p.m. is negligible making it far from easy to start from rest in the very high first gear. The car refused to start on a 1 in 3 test hill and even on a 1 in 4 it found the greatest difficulty. The clutch takes up the drive quickly, and since there is little power to drive the car forward, and not enough "feel" in the very light pedal, it is difficult to sustain engine speed on rapid standing start take-offs. Once the car is under way, however, acceleration is swift right up to well over 100 m.p.h. and only when the red line on the tachometer at 6,500 r.p.m. (108 m.p.h.) is approached in top gear, does it tail off appreciably. There is some mechanical noise from the chain drive to the overhead camshaft, but the engine is reasonably quiet, and the exhaust a subdued rasp, not unpleasant, and not loud enough to attract the wrong people. Cold starts were easy, and the only trace of temperament was during sustained high speed runs when there was intermittent pre-ignition which took a minute or two to resolve itself, a fault which a harder grade of plug would probably put right. Remarkably cool running was a feature of its performance, except in traffic when, owing, we suspect, to the thermostatically controlled electric fan not working, the engine boiled.

The Elite is all the better for its new German ZF close ratio gearbox with synchromesh on all four speeds. It is not a quiet box; all the speeds with the exception of top whine quite loudly, but gear-changing is so smooth and positive that it works like a rather large switch. Lever travel is short and the powerful synchromesh is extremely difficult to beat. On a car of this character, the close ratios were particularly appreciated and more than compensated for the difficult take-off from rest. The only serious failings of the transmission seemed to be rather a lot of snatch from the short shafts with their lack of "cushioning" effect and a thud from the final drive casing when the clutch was let in. Otherwise, making the free use of the gearbox necessary to get the best out of such a small, fast-revving engine was a pleasure.

Good ride

NOT unexpectedly, the Lotus Elite turned out to have fine road-holding. The Pirelli Cinturas with which the car is now equipped are quiet running and complement its ability to corner fast with only a little, well-damped roll. What was more surprising, however, was the exemplary ride. Large, quite abrupt bumps at near maximum speed were marked by a gentle rise of the car on its springs and as gentle a fall. Such movements were damped out immediately and pitching was almost totally absent. At low speeds, very rough roads produced some jolting and loud thumps as the suspension did its work, and the steering kicked back enough to make driving on such surfaces uncomfortable. The Elite is a splendid example of a successful compromise between the firmness associated with good roadholding and the riding comfort of a large, touring car.

The boot is very deep and holds a surprising amount of luggage. There is no support for the lid which has to be held open with one hand while luggage is loaded.

Special Equipment Elites have two 1½-in. S.U. carburetters on the Coventry Climax engine. There is no prop to hold the bonnet open, an annoyance when topping up with oil.

63

The steering is light and smooth, but lacking in the feel expected of rack-and-pinion, probably because there is very little castor action; this has the advantage that the steering is extremely light which makes an arm's length driving position fully practical.

On Brands Hatch circuit, the near-neutral steering characteristics were exploited to direct the car accurately on a given "line" which it would hold with a fine disregard of the speed it was doing. Wheel adhesion is excellent and the car remained stable and safe, with plenty of warning of the limit being reached. Sensitive to tyre pressures, the Elite seemed to handle best at inflations above the manufacturer's recommendations and its performance at Brands indicated that even higher pressures than those we used most of the time would sometimes be desirable provided the same front/rear relationship was maintained. In the wet, the car could be handled with a great deal of confidence, there being no particular necessity to exercise restraint with the throttle, even in the lower gears as is usually the case with small, light, powerful cars, the independent rear suspension distributing the power evenly and limiting wheelspin.

The brakes needed a very high pedal pressure indeed, but they also gave the driver confidence in wet or dry, and needed a very hearty push to produce wheel locking in either condition. Fade was never experienced during our test and although the rear discs are inboard, cooling does not appear to present any problem. The handbrake, however, is unsatisfactory and the hardest application failed to hold the car on a 1 in 4 gradient.

Contemporary decor

ALTHOUGH the Elite feels rather "high-waisted" and the bottoms of the windows unfashionably high in relation to the driver's eye level, the appearance of the inside of the car is unquestionably up-to-date. There is a good deal of non-reflective black and the instruments are laid out sensibly and conveniently. Lighting is good, switchgear well-placed and for such a compact car, there are lots of places to put things. The big doors have useful pockets, there is a parcel shelf and behind the long-backed seats the spare wheel cover has a large shelf with a lip. Comfort is catered for by a powerful heater but vision is neglected with no demister and two-speed windscreen wipers which lift off the screen at speed. Carpets are good and well fitted.

Further side-effects of the excellent aerodynamic shape of the Elite are noticed in the small amount of throttle needed to maintain it at speeds very near its maximum. Once the car is travelling fast, the throttle can be backed off and the speed will remain unaltered, giving very economical high-speed cruising, which is just as well with a fuel tank holding only 6½ gallons. With the side windows completely removed (the shape of the doors would not accommodate winding windows), little or no draught comes in, even when travelling very quickly. A much less pleasant feature is its sensitivity to cross winds; slight gusts making quite substantial steering corrections necessary, and limiting the usable speed on windy days.

The Elite's driving position is responsible for a large part of its appeal. The driver is held firmly in place so that he can enjoy this car on the twisty roads which are its natural habitat. They improve with occupation, if anything, the long backs giving ample support all the way up the spine and the broad, flat cushions prevent rolling on corners. The steering wheel is nearly vertical, and the driver has lots of elbow room. There is plenty of seat adjustment for the tall and the gear lever is well placed on top of the substantial tunnel between the seats. The handbrake is less than ideal, a pistol grip lever under the facia being out of place in this car. The pedals are well placed for heel-and-toe changes, but the test car's throttle had a very sticky action until the linkage was lubricated properly.

Vibration

A SERIOUS shortcoming of the Elite is the amount of noise and vibration at most speeds. At about 30 m.p.h., for example, there was a low-frequency shake. Around 100 m.p.h. a quite different vibration took over and made the very air inside the car reverberate. Wind noise was low, but near maximum the occupants were getting a full vibro massage and found difficulty making themselves heard shouting to each other. Road noise is transmitted up the back suspension struts into the body shell where it seems to be magnified into an unpleasant and most tiresome rumble which goes on most of the time. With the thumps from the suspension, road rumble, chain rattle and valve clatter from the engine, gear whine and exhaust noise, the Elite is by no means quiet. In fact the noise and vibration spoil it as a long distance tourer; a task for which it would otherwise be ideal, luggage and passenger accommodation being admirable, and economy and performance splendid.

A number of trivial faults spoilt our test of the Elite, but on the whole the car performed well. It is a most satisfying car to drive and improves with acquaintance, once one becomes used to such a small, light car having such a remarkable performance coupled to extreme docility and such superlative handling. It sets a standard, both on and off the circuits, by which most small G.T. cars are judged.

The small bumpers are close to the body and protrusions are reduced to the minimum. The snap-action fuel filler is well-placed for easy filling but is too small to take full delivery from most pumps.

Coachwork and Equipment

Starting handle	None
Battery mounting	In boot
Jack	Screw scissors type
Jacking points:	Two points under each side of body
Standard tool kit:	Jack and handle, three spanners, adjustable spanner, sparking plug spanner, pliers, screwdriver, wheel nut hammer.
Exterior lights:	Two headlamps, two side lamps, two stop/tail, two number plate lamps.
Number of electrical fuses:	One plus one circuit breaker.
Sun visors	None
Direction indicators:	Self-cancelling time-switch flashers.
Windscreen wipers	2 speed Electric, self parking
Windscreen washers:	Foot-operated plunger type
Instruments:	Speedometer with total and decimal trip mileage recorders, tachometer, oil pressure gauge, water thermometer, ammeter, fuel gauge
Warning lights:	Dynamo charge, direction indicators, main beam.
Locks: with ignition key:	Ignition/starter switch, driver's door, luggage locker, no other keys.
Glove lockers	None
Map pockets	Inside doors
Parcel shelves	Below facia, behind seats
Ashtrays	Two in doors
Cigar lighters	None
Interior lights	One in roof
Interior heater	Optional extra
Car radio	None offered
Extras available:	Seat belts, racing equipment
Upholstery material	P.v.c.
Floor covering	Carpets
Exterior colours standardized:	Five: Silver roof on special equipment model.

Maintenance

Sump	8 pints, S.A.E. 40/50 summer, 20/30 winter
Gearbox	1 pint, S.A.E. 20/30 touring, 40/50 racing
Rear axle	1⅜ pints, S.A.E. 90
Steering gear lubricant	Grease
Cooling system capacity	12 pints (1 drain plug)
Chassis lubrication	By grease gun every 1,500 miles to 8 points
Ignition timing	2-3° b.t.d.c. static
Contact breaker gap	0.014-0.016 in.
Sparking plug type	Champion N3
Sparking plug gap	0.018 in.
Valve timing:	Inlet open 12° b.t.d.c. and closes 56° a.b.d.c. Exhaust opens 56° b.b.d.c. and closes 12° a.t.d.c.
Tappet clearances (cold):	Inlet .004 in Exhaust .004 in.
Front wheel toe-in	$\tfrac{1}{16}$ in. to $\tfrac{1}{8}$ in.
Camber angle	1¼° to 1½°
Castor angle	7°
Steering swivel pin inclination	9°
Tyre pressures:	Front 21 lb., rear 25 lb. (high speeds); front 19 lb., rear 23 lb. (touring).
Brake fluid	Girling crimson
Battery type and capacity	Exide 51 amp.-hr.

1 LOTUS ELITE S/E

LOTUS UPSET THE APPLE-CART by giving us an Elite to play with for a week; it has spoilt the taste of bread-and-butter cars. And trying out bread-and-butter cars is what we are here for.

Let's say at the start that the Elite is well-named. It offers a kind of motoring that John Citizen will never enjoy in our time, and probably doesn't even know exists.

There are things wrong with the car, since it was designed and built by fallible human beings. But for our money we will take the noise and the vibration and the other little quibbles for the sake of the exhilaration and it's-good-to-be-alive feeling it gives the driver.

To begin with it looks beautiful and purposeful. You can spend an enjoyable 10 minutes just contemplating this rare product of the coachbuilder's art. Coachbuilder? Certainly. Maybe he doesn't work in wood and aluminium any more, but there's thought and loving care and practical know-how in every sinuous curve of the glass-fibre envelope that hides the Elite's rude mechanicals.

And that is what the package is for. To cheat the wind, cover up the works, and provide a ridiculously meagre petrol consumption for an express carriage. By way of good measure the Elite also offers more luggage space than most sports cars, but it is a pure two-seater unless you have a micro-midget baby in a matching carry-cot to go in the back.

Drop inside, and you have room for two and their preferably soft-packed luggage. Those curious suspension units and the spare-wheel moulding behind the seats fill up the rear, except for some space on top for coats, maps and not-too-bulky odds and ends. In the tail there is reasonable room for small suitcases and the like in a

65

lockable boot which lacks a prop for when you are loading. But being glass-fibre like the rest of the bodyshell it is light and no hardship to hold up for short stretches.

When you're ready for the road the driving position is the nearest thing to perfection we have found. With the seat well back the small-diameter woodrim wheel is just right for a straight-arm man and the pedals come to the feet and not the knees, as on some cars we could mention but won't.

Just to prove it *can* be done, the Lotus designers have not only put the wheel, the seat and the pedals in the right places; they have the gear-lever where you don't need to be an orang-outang to reach it, nor to have eyes in the side of your head to see the instruments which sit readily visible between the two top spokes of the steering wheel, only a nod of the head from the road.

They had to spoil it somehow. The fly in the ointment is a rather nasty sort of handbrake lever which hides under the dash and has to be wrenched over the ratchet, unless we just had a bad one. It is a pull-and-twist affair which is intrinsically bad anyway. Why not a simple lever of the good old-fashioned kind, fly-off for preference?

The Coventry-Climax engine on our car, a new one barely run-in, ticked over from 1000 to 1200 rpm but the clutch could be engaged without crunch or creep with minimum motion of the pedal. Once on the move the world is your oyster. Think of your normal time on a journey, cut in two, and that is somewhere about Lotus-time.

The red line on the rev counter said 6500 was the limit, and using this the gear changes come up at 50, 70 and 90-plus. This makes other traffic rather a nuisance, except that the small size and handiness of the Elite demand only a small hole and a brief stretch of clear road to overtake.

One thing puzzled us. The rapid acceleration, Gran Turismo road-holding and giant-hand disc brakes of the Lotus make it surely one of the safest things on four wheels. Yet insurance men scowled and shifted in their seats when we asked for cover. Why, in Heaven's name? If you get into trouble with this car don't blame the designer. It's plain pilot error.

Before some calculus-cowboy does quick sums and takes us up on the gear speeds, let us say that the Elite we drove was the special-equipment model with the ZF close-ratio box, which comes in place of the cooking model's MG cogs. S/E also brings synchro on all four, fresh-air heater, quick-release filler cap, Cintura (in our case) or X tyres and a bunch-of-bananas exhaust system.

In the first few miles we found out that you just don't corner the Elite on trailing throttle: that way lies the breaker's yard. We went into a sharpish bend that turned out to be even sharper, lifted off to help the tail round. But the front went instead of the back, and there we were — a sitting target for a double-decker or similar.

This is a lesson quickly learned. If there is more corner than you allowed for, just cross the hands, tread on the loud pedal, and all will be well. Honest. This business of being under power calls for a quick cog snatch from time to time, and there's no problem there. Gears are all in one box, very adjacent, and to coin a phrase come readily to hand.

Once you have learned your lesson it's heigh-ho for the open road. When the West Country traffic is thick on a Saturday morning there's not much call for top gear and plenty of fun to be had using the box. When lesser vehicles slow for some fiddling obstruction like a hairpin you drop one cog and press on in the supreme confidence that this earthbound satellite provides at all times.

The Lotus can be driven in many ways. Not much happens under about 3000 rpm and if in the mood the driver may purr along on a whiff of throttle, making early upchanges and settling for cruising at 70 or so. But this is not what people plan to do when they buy a 13-cwt car with 80 or more wild horses on tap. No. At about 4500

FACTS LOTUS ELITE
How much? See text
How fast? 118 mph. Acceleration: 0–30 2.5 sec, 0–40 4.0 sec, 0–50 6.4 sec, 0–60 10.0 sec, 0–70 14.1 sec, 0–80 19.2 sec, 0–90 24.4 sec, 0–100 31.0 sec
How thirsty? 31 miles per gallon (super)
How heavy? 1456 lb
Mechanical details: Four-cylinder in-line Coventry-Climax light alloy water-cooled single ohc engine at front driving rear wheels; 1216 cc developing 80 bhp; disc brakes; all-independent suspension by coils and wishbones (front) and Chapman-strut
Greasing: 14 points every 1500 miles
Oil change: Every 1500 miles
Body: Two-seat two-door closed coupe in glass-fibre; rear boot; detachable windows

---

the power comes in with a harsh snarl and you feel drum-beats on the wind outside.

The exhaust begins to echo off the bare brick walls and life is worth living again. You sit back in seats really designed to hold people with arms and legs and a normal quota of fatty tissue, move the wheel the few degrees needed from time to time, and just enjoy yourself.

This question of noise must be considered. Too many decibels can be tiring, yet the Lotus can be driven for hours on end and the driver step out as fresh as a GI and ready for more. No-one has wasted much time or effort on sound-damping the Elite, so the mechanical music goes round and round with the bodyshell acting as a sounding box. Keep the speed around the 100 mph or so (which becomes a normal gait) and the other noises are lost in the wind. But at lower speeds they are more dominant and would not be tolerated in a more mundane model.

In our early hours of Lotus-lashing the speedometer cable sheared, which made accurate performance testing difficult, but we saw the needle hovering around the 120 mark whenever the road was straight. Lotus claim 120 from this special equipment model, so there's not much to quibble over.

The stopping matches the going, as was observed by the driver of an inferior sports-car who came upon us as we were applying the discs hard. He went by sideways and motored away backwards, fortunately unharmed but with another gem of knowledge under his cap.

One irritation was the need to make frequent stops at filling stations, caused by the small size of the tank and the rate at which the Lotus eats the miles. Six-and-a-half gallons is too small for a grand tourer, and the Lotus answer at £23 10s for a so-called longrange ten-and-a-half gallon tank seems a cheeky one.

When we stopped for the night we locked the Lotus up, but when we wanted a forgotten garment from inside the cockpit the door stayed locked. Fortunately (or unfortunately) Elite windows are light Perspex and could be persuaded, but the door stayed obstinately locked throughout the test. The detachable windows can be whipped in or out in a few seconds, but are not the most practical way of weatherproofing. Something of a hangover from the old celluloid sidescreen, we suppose.

Our car was a joy on the open road but a misery in traffic. The heat gauge went right off the clock, petrol vapourised by the high under-bonnet temperature made a smell, and the engine stalled at the traffic lights. Colin Chapman's

OVER ▶

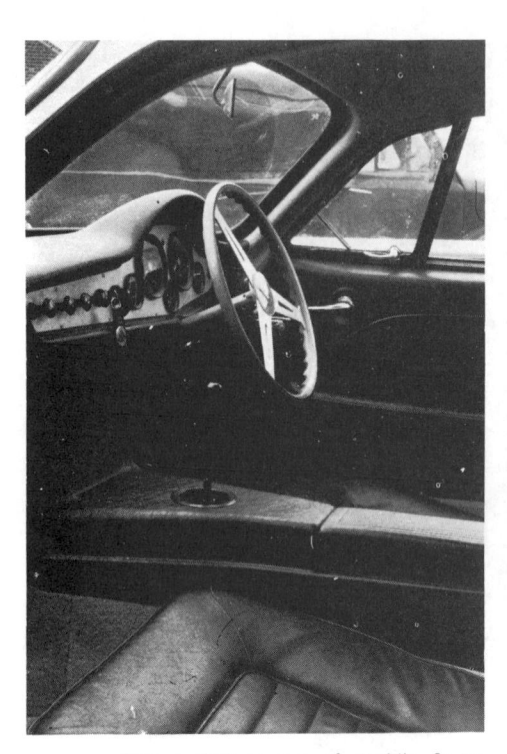

Elite cockpit is one of world's finest, offers superb contoured seats, ideally placed controls, plus full instrumentation

All-independent suspension goes some way toward explaining unique Lotus roadholding. This is final drive unit

Climax, but for how long? Test car's twin-carburettor, single-cam engine fits neatly in moderately accessible compartment

man said this was not the way Elites usually behaved, and was probably due to a faulty setting of the thermostatically controlled electric cooling fan.

We also had trouble with the first start each day, having to churn and churn like a milkmaid. This again we were told should not happen. We suspect a faulty valve in the mechanical pump allowed the fuel lines to drain overnight.

All in all we would accept the Lotus with its faults for the sake of the sheer pleasure to be had from driving it. Bought in kit form—the way 90 per cent are—it costs £1375, or £1495 with special equipment. Anyone with reasonable mechanical skill is said to be able to assemble the kit in 24 hours, and the makers will then check it over.

Ready-made prices shoot up to £1891 and £2056 with tax. All sorts of extra equipment can be added if you have the money to spend, like Weber 40/DCO/2 carburettors at £80, brake servo at £27, oil cooler at £31 and other goodies.

The £18 fresh-air heater was fitted on the test car and worked well enough in mild autumn weather. The perspex sidescreens could be left out and the crew still be comfortable. If a draught arose replacing the passenger's window only did the trick.

On Road and Track with the HOBBS MECHA-MATIC TRANSMISSION LOTUS ELITE

SOME competition cars which are outwardly similar to others in their class win far more races than their competitors. This is very often due to the car having a superior driver or a more powerful engine, or better "sorted" handling, but very seldom is the success of a car attributed in part to its gearbox. When the gearbox is an automatic one then the chances of success appear to be remote indeed, however good the car or driver may be. In the case of David Hobbs and his Lotus Elite fitted with the Hobbs Mecha-Matic automatic transmission one can merely let the results speak for themselves. Apart from successes in International competition during 1961 and 1962 he has won numerous Club and National events, including a run of 14 successive wins during 1961.

The history of the red and blue Elite 5649 UE began, as far as David Hobbs was concerned, in November 1960 when the car was purchased from the Chequered Flag sports-car garage in Chiswick. The engine was removed and sent to Lotus for Cosworth to bring up to stage III tune, and when it was returned it was giving 108 b.h.p. Meanwhile a standard Hobbs 1015 automatic gearbox as made for the Ford Anglia was prepared for the Elite. The main preparation consisted of removal of the safety jets which limit the r.p.m. at which downward changes occur. On the standard box this ensures that the engine cannot be over-revved, but in competition it is sometimes necessary to get into a low gear really quickly in the event of brake failure or similar contingency. The tow start valve was also removed. Other work on the Elite consisted of stripping out all non-essential trim and sound deadening material, including some of the double skinning of the glass-fibre chassis/body unit in parts where stresses are not high. The sidescreens and rear window were removed and replaced by Perspex ones, although the laminated windscreen was of course retained. The seats were removed and replaced by lightweight glass-fibre seats as used in the Lotus Formula Junior cars, these being fixed in one position and braced at the rear to prevent any movement. The driver's seat was fitted with a shoulder harness safety belt. The battery was removed from the boot to the floor in front of the passenger as the voltage drop along the cables gave the battery little chance of turning over the engine with its 11 : 1 compression ratio. A 15-gallon fuel tank was fitted in anticipation of long-distance racing and a large petrol filler cap was fitted. Twin S.U. electric petrol pumps fitted in the boot above the tank ensure that full fuel pressure is maintained at all times.

In the interior the speedometer was replaced by a Key Leather 9,000-r.p.m. electronic rev.-counter although the standard Smiths rev.-counter was retained. All the switches on the facia were retained but there is no heater or direction indicators and the hooter is now a pip-squeak of a thing from a Mini Minor. On the vertical face of the gearbox tunnel two temperature gauges are fitted, one for the engine oil and one for the transmission oil.

Having carried out all these modifications. David Hobbs entered the car for its first race at Mallory Park on Easter Monday, 1961, with no success at all. This was followed by a Silverstone meeting on the long circuit with no great success, the car doing around 2 min. 12 sec., against the 1 min. 55 sec. of David Buxton's Team Elite car. At the next Mallory Park meeting he was third behind a Team Elite car and Fergusson's very quick Turner. At Snetterton the following week David Hobbs thought he would be clever and put only three gallons of fuel in the 15-gallon tank for a short race, but on the corners the fuel surged away from the pumps and the engine cut out when the car was in the lead. At Brands Hatch throttle linkage trouble forced the car to retire, and at a Silverstone Club meeting the Elite indulged in a dice with Dawson's Lotus XI, Hobbs leading the XI all the way round except on the long straight into Woodcote, Dawson just holding him off at Woodcote to rob him of his first win. He was also second in several other races that day. The next event was the Silverstone International meeting where, because the car was not homologated with the F.I.A. with its automatic gearbox, it was put into the sports-car race, which was won by Stirling Moss' Lotus XIX. He held tenth place in this exalted company until a plug lead fell off, the subsequent pit stop losing him over a minute; but he still finished twelfth. The next day, at the Lords Taverners meeting at Brands Hatch, he got his first big win by defeating Les Leston in another Lotus Elite. This led to his run of 14 successive wins in G.T. races in this country.

The first foreign outing was at the Nurburgring 1,000 kilometres, where the car was protested out of the G.T. class into the sports-car category. Hobbs and co-driver Bill Pinkney limited themselves to 6,500 r.p.m. but managed to lap every other Elite except the Lumsden/Riley car, which beat them, as well as many Porsches and Alfa Romeos. They finished 20th overall and won the 1,600-c.c. sports-car class after Heini Walters' RS61 Porsche had crashed. The car had no less than seven stops for oil which lost it a lot of time. Straight from the Nurburgring Hobbs went to Oulton Park and in the wet, using Michelin "X" tyres, won three races including one for unlimited sports cars. Many other Club races came his way, including the Astley Trophy

STARK interior of the Hobbs Lotus Elite.

69

at Aintree, and at the British Empire Trophy meeting he finished fifth overall in the G.T. race, the cars that beat him being two Ferraris, a Jaguar and an Aston Martin. However, he gained no class award as he was told just before the race that the car did not comply with the rules despite being assured that it was eligible when it was entered. The next foreign trip was to Pescara for the four-hour race. In practice Hobbs and Pinkney were miles faster than the hot Alfas in their class so were moved by the organisers into the 2-litre sports-car class! Pinkney started the race and was out after 1½ laps with a rod through both sides of the block due to a big-end bolt fracture. This was repaired and the car next appeared in a Club Silverstone meeting, where a front wishbone mounting tore out of the glass-fibre body when braking for Woodcote. This is a trouble that the Elite is prone to and after it happened again David Hobbs' mechanic, Ken Taylor, fitted a bar across the car connecting the two wishbone mountings, which seems to have cured the trouble. The other Elite trouble of the chassis-mounted differential twisting and finally tearing away from the body was forestalled by more rigid location of the differential when the car was being prepared by ex-Lotus mechanic Henry Lee in the winter of 1960. The season wound up with the Clubman's Championship, in which David Hobbs finished fourth in the G.T. race despite losing a lot of oil from the gearbox.

Nineteen sixty-two started with a couple of class wins in Club meetings but as Hobbs was by now in the Peter Berry racing team driving 3.8 and E-type Jaguars he had little time left to drive the Elite, but managed a return trip to the Nurburgring, where he was *ninth* on the first lap in front of works Ferraris and Porsches. They had been put into the 2-litre sports-car class against works flat-eight Porsches. When in 12th position after 5½ hours' racing and having lapped all other Elites the cylinder head cracked due to overheating and they were forced to retire. In this race Hobbs and co-driver Richard Attwood were using 7,000 r.p.m., as against the 6,500 r.p.m. of the previous year. The rest of the season saw little activity for the Elite as its owner was busy in bigger cars, but it was taken out to Clermont-Ferrand for the three-hour race. Unfortunately, when lying seventh overall, a stone became wedged between pad and disc of one of the front brakes, with the result that when the brakes were applied the wheel locked and the car left the road and hit a bank rather hard. The resultant severe damage required a completely new bodyshell, although Lotus managed to rustle up one which had been involved in a minor prang, so all the bits and pieces were transferred to this car. This brings us practically up to date; the last outing of the car was at the Guild of Motoring Writers' test day at Goodwood, when dozens of journalists thrashed the car round Goodwood, completely ignoring the owner's pleas not to exceed 6,500 r.p.m.!

This short trip whetted our appetite for more experience with this car and David Hobbs readily agreed to our suggestion to carry out a test on the Elite. On a sunny but cold day in early November we met at Silverstone where the car was to be put through its paces on the Club circuit.

The Elite has a normal looking gear-lever in the usual position but this moves only in a fore and aft plane, and a plate on the gearbox tunnel shows the gear positions. Reverse is obtained by lifting the gearknob and pushing the lever right forward, with neutral being selected by bringing the lever back one notch. The forward gears are then selected by bringing the lever back one further notch for each higher gear. The rearmost position of the

SLEEK.—*The profile of the Elite is still one of the most handsome on road or track.*

HOT.—*The engine is tuned to Stage III with maximum power of 108 b.h.p.*

lever selects top gear and is also the position for fully automatic operation of the gearbox. For road use the lever can be left in this position and the gearbox will do all the work. On a light throttle upward changes occur at 15, 30 and 53 m.p.h., but the Hobbs box can have its gear-change speeds adjusted to suit the customer. The 53 m.p.h. at which this particular car changes into top is rather higher than would normally be the case but this is done on the racing Elite to eliminate any chance of plug wetting due to lugging in top gear at low speeds. On the circuit David Hobbs selects the gears manually so that he can hold any gear to any revs he likes, and in moments of stress in Club races he has seen 8,000 r.p.m. in top, although he does not exceed 7,500 r.p.m. normally.

Having warmed the car up on soft plugs a set of hard plugs were inserted, which promptly refused to fire more than two cylinders, so the soft plugs were re-inserted, which meant that high revs could not be indulged in for long periods as the points have a habit of disintegrating. Acceleration and top speed is also adversely affected. David Hobbs took me round for a few laps to show me his technique with the automatic box. Racing starts are made by holding 5,000 r.p.m. on the tachometer with the box in neutral, then pulling the lever back to first gear. After a momentary pause the car screams away up to maximum revs in first, the lever being pulled back into the next higher gear with the throttle flat on the floor, only a mild jerk and a drop in revs indicating that the box has changed gear. Procedure for downward changes follows normal practice to a certain extent as David Hobbs brakes to the required speed, pushes the lever to the lower gear then blips the throttle with his heel, which has the effect of engaging the lower gear slightly quicker than if he waited for it to change itself. Having acquainted myself with the technique, I donned David Hobbs' crash helmet and ventured out on to the Club circuit. Having driven several Elites already I was fully aware of the cornering capabilities of the car, but it is always a pleasure to renew acquaintance with such superb handling. The stiffer suspension gives virtual roll-free cornering and the Dunlop R5 racing tyres are reluctant to break away unless pressed extremely

Continued

ACCELERATION FROM REST	
0– 30 m.p.h.	4.3 sec.
0– 40 m.p.h.	5.9 sec.
0– 50 m.p.h.	7.8 sec.
0– 60 m.p.h.	9.9 sec.
0– 70 m.p.h.	12.7 sec.
0– 80 m.p.h.	16.2 sec.
0– 90 m.p.h.	21.9 sec.
0–100 m.p.h.	29.1 sec.

SPEEDS IN THE GEARS (m.p.h.)		
	Manual	Automatic
First	44	15
Second	62	30
Third	82	53
Top	106	106

HOBBS LOTUS ELITE—continued

hard indeed. On open bends handling is near neutral but at Becketts some understeer becomes apparent, although David Hobbs creates oversteer by twitching the car into corners slightly faster than they can safely be taken! In the Elite Copse corner is taken in 3rd, at around 6,000 r.p.m., Maggotts in top at maximum revs, the drift to the right of the track being halted in time to swing over and brake really hard for Becketts, which is taken in 2nd gear. Maximum revs in top are reached halfway down the straight, my self-imposed maximum on the first outing being 7,000 r.p.m. For Woodcote the brakes go hard on just before the 100-yard board and 3rd gear selected, although David Hobbs goes down to 2nd to give better acceleration out of the corner. However by using 2nd the revs are dangerously high, so as I was due to take the car away for road-test afterwards I settled for 3rd gear. After about 10 laps in this exhilarating car I pulled in and handed over to David Hobbs, who soon settled down and lapped in 1 min. 13.1 sec. in five laps or so. After his spell I got back in and by using 7,500 r.p.m. my previous best time of 1 min. 18 sec. was lowered first to 1 min. 16.3 sec., then to 1 min. 15 sec., and finally to 1 min. 14.1 sec. Since the best Elite time ever recorded on the Club circuit is around 1 min. 11 sec. and that by a car with a special body/chassis unit which is so light that, in David Hobbs' words, "I don't know how it stays together," it can be seen that the combination of this well-maintained Elite and the Hobbs transmission form a combination which a driver of average ability can take round a circuit in very creditable times. Far from detracting from the performance the Hobbs box obviously consumes very little power and the time saved by eliminating the clutch is reflected in the excellent lap times.

Having enjoyed the car on the circuit we prepared to enjoy it on the road. The preparation consisted of taking off the short exhaust system and replacing it with a longer pipe with a small silencer which reduces the noise level to acceptable limits for town use but also reduces performance somewhat. To give a softer ride on the road the wheels were changed for a set fitted with Michelin "X" tyres. As David Hobbs has proved, they also give better handling in the wet on the circuit but cannot match racing tyres in the dry.

Without a crash helmet the noise inside the Elite is cacophonous, the noises of gearbox, final drive, engine, suspension and exhaust combining to make a fearsome racket at high speed. However, outside the noise is not quite so apparent and the Elite could be trickled past policemen on point duty without attracting their attention at all. Many road users recognised the car and some tried to keep up with it, but the superb handling and useful turn of speed allowed it to pass virtually anything with ease. We took a set of performance figures just to see what a well-tuned Elite would do and despite our inability to make a good getaway from rest it reached 80 m.p.h. in 16.2 sec., which is not hanging around by any standard, while 100 m.p.h. is available in under half a minute from a standstill. In deference to the soft plugs which are used on the road we did not exceed 7,500 r.p.m. and at this figure we obtained speeds of 44, 62, 82 and 106 m.p.h. in the gears. A higher axle ratio would give more impressive speeds in the gears but the present ratio is most suitable for short circuit work.

The gearbox behaved in a most impressive fashion in conditions ranging from flat-out cruising to trickling through town traffic. The engine temperature never exceeded 90° C. at any time, while the transmission oil temperature never went above 70° C. even during performance testing. By leaving the lever in "Automatic" the car could be driven in a most relaxed manner, the only snag on this particular installation being that top gear will not engage until 5,000 r.p.m. is reached. This means that in 40-m.p.h. limits the car is in 3rd gear all the time. Changes take place with only a moderate jerk both up and down, and on engines with a better torque curve these jerks are claimed to be much less severe. Of course the great advantage that this transmission has over other automatic types is that any gear can be selected at will and held until the lever is moved again. Putting the lever in a particular gear does not mean that it sticks in this gear, and if speed drops to such an extent that a lower gear is necessary it will change automatically, so that the lever can be left in the 3rd-gear position and the car allowed to come to rest, then accelerate away again until 3rd gear is attained once more.

The fitting of the Hobbs gearbox in a racing Elite is probably one of the most difficult applications that an automatic transmission can have. I am happy to report that it is completely successful in every way. Fitted in a more conventional car it must be many people's idea of the perfect car. Stirling Moss thinks so, for his personal Elite is fitted with a Hobbs gearbox.—M. L. T.

Continued from page 15

this type of car, has a vast amount of space available for luggage. Here again the finish of the internal panels is comparable with the exterior.

The all-independent suspension smooths out road irregularities at speed but was a trifle harsh at low speeds on the test car due to it being fitted with non-standard "competition" rate springs. With driver and passenger aboard the acceleration is fantastic, 100 m.p.h. being easily obtainable on the Thame-Bicester road, although the presence of a lorry around every bend made the quest for the obvious 115 m.p.h. maximum a "dicey" procedure on what was after all a brief trial run.

Swinging the Elite around the Buckinghamshire bends at high speed was an unusually invigorating experience, the road ribboning out ahead of the low-slung bonnet, and the exhaust note rising to a crisp crackle (a non-standard system on the test car) as the gears were slipped in with typical M.G. ease. The car becomes alive in the hands as the revs build up—an easy 85 in third and still plenty of kick-in-the-back as top gear is snatched. The high-geared steering (two turns from lock to lock) was a joy, only a slight movement being necessary to swoop round the bends—bends which became corners at Lotus Elite speeds, but there was always the feeling of security from the Girling disc brakes. Steering lock is impressive with a turning circle of less than 35 feet.

Summed up there can be few "road" cars which could live with the Elite on a short or long journey under modern traffic conditions—or even on deserted roads! It is the modern sports car personnified—room for two people in great comfort, complete weather protection, ample luggage space, good looks, and near-racing car performance. Coupled with these features is a relatively low fuel consumption. With its excellent aerodynamic shape and high gearing it would undoubtedly return figures of around 40 m.p.g. if driven at reasonable—but not low speeds.—D.A. ★

SPECIFICATION

ENGINE
Four-cylinder, Coventry-Climax single o.h.c. Capacity: 1,220 c.c., Bore: 76.2 mm., Stroke: 66.6 mm. Compression ratio 8.5 : 1. Max. Power: 75 b.h.p. at 6,100 r.p.m. Single 1½ in S.U. carburetter.

GEARBOX
M.G.A. four-speed close ratio, synchromesh on 2nd, 3rd and top. Ratios: 1st, 3.67 : 1; 2nd, 2.20 : 1; 3rd, 1.32 : 1; Top, 1 : 11. Reverse, 3.67 : 1.

FINAL DRIVE
Hypoid final drive unit. Standard ratio 4.875 : 1. Optional available ratios: 5.375, 5.125, 4.875, 4.55, 4.22, 3.89, 3.73 : 1.

BRAKES
Ultra light hydraulically operated 9½-in. Girling disc brakes, outboard at the front and inboard at the rear. Handbrake operating rear calipers through cables.

TYRES AND WHEELS
High performance 490 × 15 Firestone nylon tyres (Michelin X optional) front and rear fitted on knock-on wire wheels with identical rims front and rear.

SUSPENSION
Front: Independent transverse wishbone incorporating a roll bar with combined coil spring-damper units. Rear: Independent Chapman strut system incorporating combined coil spring-damper units and including double articulated drive shaft giving lateral location as developed on the Lotus F1 and F2.

TANK CAPACITY
Side tanks moulded in wings giving total capacity of approximately 7 gallons.

STEERING
Lightweight rack and pinion. Adjustable steering column.

DIMENSIONS
Wheelbase: 7 ft. 4 in. Track: F. and R. 3 ft. 11 in. Overall Length: 12 ft. Overall Width: 4 ft. 10 in. Height to roof: 3 ft. 10 in. Weight dry! 10¾ cwt.

PRICE
Basic: £1,300. Total including tax: £1,951 7s.

LOTUS ELITE

*Noisy, but nice; the smallest, prettiest
and most intriguing GT car in a decade*

LONGTIME READERS of *Road & Track* will need no introduction to the Lotus Elite, but for those who are not familiar with the model, a few words of explanation are in order.

Introduced at the Earls Court Show in London, in the fall of 1957, the Elite represented the most outstanding esthetic automotive design to come out of England since the MG-TC. And in many ways it surpassed even that venerable example of automotive architecture, for it made a complete break with British "traditional" design.

Beneath the esthetically beautiful body, however, the car was even more interesting. The construction was an integral unit body/chassis made entirely of fiber glass—reinforced plastic; there was no metal used in the structure—a feature that was, and is, unique to the Elite.

The standard engine, and in fact the only one available in the car, was the single overhead camshaft 4-cyl Coventry Climax FWE unit. This was the engine that started life as a stationary "fire pump" engine but was rapidly becoming the most popular engine for sports/racing cars, particularly the Lotus Eleven. Of 1216-cc displacement, it delivered 75 bhp at 6100 rpm with 70 ft/lb of torque at 3300 rpm. With a curb weight of 1420 lb, the Elite promised very satisfactory performance.

Naturally, every sports car enthusiast wanted to get his hands on an Elite but production problems, coupled with local demand in England, prevented *Road & Track* from testing the model until January 1960. Now, over three years later, the Elite looks the same, but changes have been made and we feel a re-evaluation is in order.

Instead of one Elite, we started home with two of them, almost identical except for color—one red and one white. The white car had the MG-A gearbox and Michelin tires; the red car had the German ZF all-synchromesh gearbox and Pirelli Cinturos. Both had left-hand drive (the car is available with either left- or right-hand drive).

Threading our way through evening traffic, the Elite seemed almost too small to be involved in this river of metal. The domestic cars alongside looked even larger than they usually do, and trucks gave us moments of panic as the rotating hub nuts flashed by at our eye level. We thought to ourselves, let's get out of here.

And out we got. At the first opportunity we took to the back streets and headed for the freeway. The 1216-cc engine pulled smoothly and apparently effortlessly. The ZF 4-speed all-synchromesh gearbox worked beautifully—shifting up or down with a minimum of motion and an ease that made gaining familiarity with the car all that much quicker.

Up onto the freeway and into the thinner, and faster moving, traffic, where the Elite's sensitive rack and pinion steering made itself even more apparent than we had noticed at the start. Changing lanes took only the thought to accomplish. The approach of yon idiot needed only a touch on the brake pedal or a minute flick of the steering wheel to avoid danger, and we soon lost our early tendency to over-correct.

Cruising along a relatively little used portion of a new freeway gave us the leisure to examine the livability of the interior. Both driver and passenger have plenty of room, the seats are extremely comfortable and the controls really *do* fall readily to hand. The seating/pedal/hand-control layout was obviously designed by someone accustomed to high speed driving—although it appeared the designer must have been more interested in where he was going than where he'd been, because vision to the rear was somewhat hampered by the lack of outside rearview mirrors. The interior mirror just didn't take in enough area to suit us.

The moderately instrumented panel is functional in the truest sense and contains a speedometer, tachometer (both large and easily read), water temperature, oil pressure, ammeter and fuel level gauges, all mounted directly in front of the driver. A rheostat controls the lighting on the circular instruments (white numbers on black) for night driving. The panel cowling prevents any windshield reflection from the instrument lights and the matte black finish on the cowl assured no daytime reflections in the windshield.

At a comfortable cruising speed, a drumming developed inside the cab that was extremely annoying. We changed speed and it diminished temporarily, only to return as the speed leveled off. During the 10 days we had the two cars this droning persisted in both, and in talking to others who've driven the Elite it seems to be peculiar to the breed. Resonance—from engine, exhaust and road vibrations—set up in unit body/chassis construction is a difficult problem to overcome, and auto manufacturers have spent millions trying. It would seem that a panel is a panel is a panel, whether steel or glass, but regardless of the reason and/or the attempted cure, the Elite drones on. The exhaust, even though rather lightly muffled, was not particularly obtrusive, but the panel resonance definitely was.

Taking a day off from the Dreaded Yard Work on the

Comfort and space that would do credit to many a larger car.

Old Homestead, we spent a Sunday putting miles on the cars, from sea level to above 6500 ft. No baggage was carried, but it wouldn't have made much difference; the car went up the mountain with no strain and, where traffic conditions permitted, could be driven in a way which would have sent lesser cars off the road with tires squealing.

As the weather was on the cold side, the lack of roll-down windows was not lamented. In more pleasant temperature zones, the need to remove the window from the door would prove to be a bother—but a door too thin to allow roll-down window mechanism is one of the penalties paid to get sufficient elbow room in a mighty small car. Removing the window takes only a matter of seconds—just a twist of the catch and out it comes, but the window is either all shut or all open. There are windwings and they provided adequate ventilation during our time spent with the cars, but roll-down windows would have been nice.

When we got to the acceleration part of our test procedure we tried out both cars and discovered that the BMC transmission-equipped model was faster off the line. The ZF gearbox carries close-ratio gears (2.53, 1.71, 1.23 and 1:1 on the ZF—3.67, 2.20, 1.32 and 1:1 in the BMC unit) and would bog down at the start, because with two men aboard the car simply would not spin its rear wheels. The BMC-equipped car, on the other hand, *would* break loose, allowing the engine to stay nearer its power peak and consequently allow a better start.

All on the staff who drove both cars still preferred the ZF-equipped version and would not consider the Elite without this unit installed.

The Elite's brakes are Girling discs, 9.5 in. dia., all around; outboard at the front, inboard at the rear, with an underdash-mounted handbrake lever operating the rear brakes through a cable. As we say about disc brakes so often—they worked magnificently and stopped the car straight and fast each time with no drama.

Suspension is independent all around with unequal length A-arms in front—the forward part of the upper A-arms being formed by the stabilizer bar—and Colin Chapman-designed strut and trailing arms in the rear. Lateral location of the rear wheels is provided by the non-splined half shafts. Springing is by integral coil spring/telescopic damper units.

The Lotus Elite shares very little in common with other cars,

The heart of the car—the Coventry Climax FWE sohc engine.

Not much room, but what's there is usable. Battery is at left.

LOTUS ELITE

but we did notice a peculiar feeling while negotiating corners at near maximum speed; a sensation not unlike other cars with 4-wheel independent suspension. A sort of "walking" condition was evident both in the corner and for 50 yards or so after leaving the corner.

We never reached the point where we felt really ill at ease, but the car does take some getting used to. Lotus devotees claim that the Elite will out-corner anything on [4] wheels, and they may be right. We're not going out on a limb at this point and take sides but, admittedly, the car does *feel* as though it would out-corner any other. And it's at its best on tight, twisty roads where maneuverability pays off.

Looking back on our test of three years ago, we find that our opinion of the Elite hasn't changed much. It's still one of the best looking GT cars ever built, it still looks like a toy but doesn't act like one, still stops on a dime and corners like a chased jack rabbit, is even more fun to drive with the ZF transmission, but still needs attention to the noise level (for touring) and price is high.

If you're the type of driver who places the pure enjoyment of driving above all else—drop into your neighborhood Lotus dealer and try one. This may very well be the car for you.

ROAD TEST
LOTUS ELITE

SCALE: 10" DIVISIONS

DIMENSIONS

Wheelbase, in.	88.2
Tread, f and r	47.0/48.2
Over-all length, in.	147.0
width	58.5
height	47.0
equivalent vol, cu ft	234
Frontal area, sq ft	15.3
Ground clearance, in.	5.0
Steering ratio, o/a	n.a.
turns, lock to lock	2.5
turning circle, ft	29.7
Hip room, front	2 x 19.2
Hip room, rear	none
Pedal to seat back	41.7
Floor to ground	7.5

CALCULATED DATA

Lb/hp (test wt)	22.8
Cu ft/ton mile	85.6
Mph/1000 rpm (4th)	16.5
Engine revs/mile	3630
Piston travel, ft/mile	1585
Rpm @ 2500 ft/min	5730
equivalent mph	95
R&T wear index	57.5

SPECIFICATIONS

List price	$4780
Curb weight, lb	1460
Test weight	1820
distribution, %	43/57
Tire size	155-15
Brake swept area	358
Engine type	4-cyl, sohc
Bore & stroke	3.0 x 2.62
Displacement, cc	1216
cu in	74.2
Compression ratio	10.0
Bhp @ rpm	80 @ 6100
equivalent mph	101
Torque, lb-ft	75 @ 4750
equivalent mph	79

GEAR RATIOS

4th (1.00)	4.23
3rd (1.23)	5.18
2nd (1.71)	7.21
1st (2.53)	10.7

SPEEDOMETER ERROR

30 mph	actual, 29.4
60 mph	58.1

PERFORMANCE

Top speed (4th), mph	115
Shifts, rpm-mph	
3rd (6400)	86
2nd (6400)	62
1st (6400)	42

FUEL CONSUMPTION

Normal range, mpg	32-38

ACCELERATION

0-30 mph, sec	4.6
0-40	6.7
0-50	9.0
0-60	11.8
0-70	15.8
0-80	20.3
0-100	34.5
Standing ¼ mile	18.3
speed at end	76

TAPLEY DATA

4th, maximum gradient, %	10.7
3rd	14.1
2nd	21.0
Total drag at 60 mph, lb	100

ENGINE SPEED IN GEARS

ACCELERATION & COASTING

LOTUS

GREAT BRITAIN

Elite Series II

ENGINE CAPACITY 74.37 cu in, 1,216 cu cm
FUEL CONSUMPTION 27.7 m/imp gal, 23.1 m/US gal, 10.2 l × 100 km
SEATS 2 **MAX SPEED** 113 mph, 181.9 km/h
PRICE list £ 1,375, total £ 1,662

ENGINE front, 4 stroke; cylinders: 4, vertical, in line; bore and stroke: 3 × 2.62 in, 76.2 × 66.6 mm; engine capacity: 74.37 cu in, 1,216 cu cm; compression ratio: 10; max power (SAE): 85 hp at 6,300 rpm; max number of engine rpm: 7,000; specific power: 69.9 hp/l; cylinder block: light alloy, wet liners; cylinder head: light alloy; crankshaft bearings: 3; valves: 2 per cylinder, overhead; camshaft: 1, overhead; lubrication: gear pump, full flow filter; lubricating system capacity: 8.60 imp pt, 10.36 US pt, 4.9 l; carburation: 2 SU type H4 horizontal carburettors; fuel feed: electric pump; cooling system: water; cooling system capacity: 12 imp pt, 14.37 US pt, 6.8 l.

TRANSMISSION driving wheels: rear; clutch: single dry plate, hydraulically controlled; gearbox: mechanical; gears: 4 + reverse; synchromesh gears: I, II, III, IV; gearbox ratios: I 2.529, II 1.709, III 1.229, IV 1, rev 2.588; gear lever: central; final drive: hypoid bevel; axle ratio: 4.20.

CHASSIS integral in plastic material; front suspension: independent, wishbones, coil springs, anti-roll bar, telescopic dampers; rear suspension: independent, trailing radius arms, transverse guide semi-axle arms, coil springs, telescopic damper struts.

STEERING rack-and-pinion; turns of steering wheel lock to lock: 2.75.

BRAKES disc (diameter 9.50 in, 241 mm).

ELECTRICAL EQUIPMENT voltage: 12 V; battery: 56 Ah; ignition distributor: Lucas; headlights: 2.

DIMENSIONS AND WEIGHT wheel base: 87.99 in, 2,235 mm; front track: 46.97 in, 1,193 mm; rear track: 46.97 in, 1,193 mm; overall length: 150 in, 3,810 mm; overall width: 57.87 in, 1,470 mm; overall height: 46.06 in, 1,170 mm; ground clearance: 7.09 in, 180 mm; dry weight: 1,513 lb, 686 kg; distribution of weight: 45% front axle, 55% rear axle; turning circle (between walls): 30.8 ft, 9.4 m; tyres: 155 × 15; fuel tank capacity: 6.5 imp gal, 7.7 US gal, 29 l.

BODY coupé in plastic material; doors: 2; seats: 2.

PERFORMANCE max speeds: 43 mph, 69.2 km/h in 1st gear; 65 mph, 104.6 km/h in 2nd gear; 88 mph, 141.7 km/h in 3rd gear; 113 mph, 181.9 km/h in 4th gear; power-weight ratio: 17.9 lb/hp, 8.1 kg/hp; carrying capacity: 353 lb, 160 kg; acceleration: standing ¼ mile 18.5 sec, 0—50 mph (0—80 km/h) 8.2 sec; speed in direct drive at 1,000 rpm: 16.7 mph, 26.9 km/h.

PRACTICAL INSTRUCTIONS fuel: 98-100 oct petrol; engine sump oil: 7.90 imp pt, 9.51 US pt, 4.5 l, SAE 20-30 (winter) 40-50 (summer), change every 3,000 miles, 4,800 km; gearbox oil: 1 imp pt, 1.27 US pt, 0.6 l, SAE 40-50, change every 6,000 miles, 9,700 km; final drive oil: 2 imp pt, 2.32 US pt, 1.1 l, SAE 90, change every 3,000 miles, 4,800 km; greasing: every 1,500 miles, 2,400 km, 8 points; tappet clearances: inlet 0.010 in, 0.25 mm, exhaust 0.010 in, 0.25 mm; valve timing: inlet opens 30° before tdc and closes 60° after bdc, exhaust opens 60° before bdc and closes 30° after tdc; tyre pressure (medium load): front 22 psi, 1.5 atm, rear 26 psi, 1.8 atm.

VARIATIONS AND OPTIONAL ACCESSORIES 3.70 4.50 4.90 axle ratios; limited slip final drive; servo brake; fuel tank capacity 10.5 imp gal, 12.7 US gal, 48 l; 4-speed mechanical gearbox (I 2.50, II 1.67, III 1.23, IV 1, rev 2.50 - I 2.53, II 1.71, III 1.23, IV 1, rev 2.59); kit form model, max power 80 or 85 hp; Elite Super 95, max power 95 hp at 7,000 rpm, max speed 120 mph, 193.2 km/h; Elite GT, max power (DIN) 104 hp at 7,200 rpm, 2 Weber 40 DCOE 2 carburettors, max speed 130 mph, 209.3 km/h, on request oil cooler.

Cars ON TEST

LOTUS ELITE

The introduction of the Lotus Elite in 1957, as Colin Chapman's first essay into the realms of high-performance road cars—as distinct from his earlier models which, although many of them could be and often were used on the road, were designed specifically with competition use in mind—presented something of a revolution in construction. The basis of the car was a stressed-skin glass-fibre shell, with independent suspension for all four wheels, a unique concept which still remains unchanged, as does the Elite's identity as one of the world's most advanced small Grand Touring cars.

The car supplied to us for test was the special equipment version, which has a more highly-tuned version of the 1,220 c.c. Coventry Climax engine and a ZF close-ratio gearbox. Despite the fact that the design of the machine has changed little in the seven years which have elapsed since its introduction, it remains one of the most attractive small cars in production, as well as having an extremely efficient aerodynamic body shape.

The only metal employed in the construction of the basic body shell is a tubular steel hoop in the roof. Despite this, however, the body/chassis is completely stiff and the car as a whole has that "all-of-a-piece" feeling. It is, of course, extremely light, and the complete car, with engine and transmission, suspension, wheels, seats, trim and fuel, oil and water weighs less than 14 cwt: it is therefore not surprising that the engine, developing 85 b.h.p., propels the machine with considerable zest. The all-independent suspension gives a firm, unusually level ride with roadholding of racing-car standards: the ZF gearbox provides a superbly well-chosen set of ratios and the steering is light and extremely positive. The brakes, if high pedal pressure is needed, nevertheless provide good stopping-power, and the interior is well-equipped and comfortable. Only in one direction does the Elite fail to achieve the same high standard: the interior noise level is extremely high, to the point of being definitely unpleasant—an extremely tiresome feature in a car which, in all other respects, is a real delight to drive, responding instantly and precisely to every whim of the driver.

As installed in the special equipment model, the engine, the four-cylinder, overhead-camshaft Coventry Climax F.W.E. unit of 76.2 mm. × 66.66 mm. bore and stroke (1,216 c.c.). Twin 1½ in. S.U. carburettors are fitted, and the compression ratio is 10 : 1, resulting in a net power output of 85 b.h.p. at 6,300 r.p.m. Bearing in mind the competition background of this power unit, it is clearly apparent that, in this state of tune, it is working well within its limits. It is extremely smooth and free-revving, and, not unexpectedly, gives its best performance when it is turning over fairly fast. There is very little power below 2,000 r.p.m., but after that there is a smooth surge right up to maximum revs. at 6,500 r.p.m. In fact, the driver is given the impression that it would go on increasing its speed for ever, but in fact, apart from the risk of damage, the power characteristics are such that the power drops off quite quickly above this speed, while some vibration was noticeable towards the upper end of the rev.-range. It was always easy to start whether hot or cold, and proved to be unusually cool-running; prolonged idling in heavy traffic naturally caused the needle of the temperature gauge to rise, but a thermostatically-controlled electric fan coped well with the situation, and at no time did the engine boil, in fact, no water was ever added to the header tank during the test period. There is little indication of the racing history behind the power unit unless one goes out to look for it, and the Elite can be driven like a sedate town carriage. The engine is sufficiently flexible to permit speeds of below 25 m.p.h. in top gear without unpleasantness, and there is no sign of temperament. It is not the quietest of engines, and there is a high level of mechanical noise, principally from the camshaft driving-chain.

Overall the car is high-geared, and this, combined with the general lack of engine power at low revolutions, has a tendency to make starting from rest a matter of some delicacy. The clutch grips firmly and progressively, with never a sign of slip, yet is surprisingly, and pleasantly, light to operate. The close-ratio ZF gearbox, featuring unbeatable synchromesh on all four forward gears, is a positive delight to use: the ratios are extremely well-spaced, providing maxima in excess of 40, 60 and 80 m.p.h., and are selected by means of a light, positive gear lever. Lever travel is short and the gate is narrow, and frequent use of the box—a necessity if the best is to be derived from the engine—is a real pleasure. The gearbox—and indeed the transmission as a whole—is rather noisy, and there is a pronounced whine on all gears

Continued

LOTUS ELITE continued

except top, while the use of the clutch is sometimes accompanied by a thud from the final drive casing. In addition, a pronounced snatch from the half-shafts, which lack cushioning, was occasionally felt.

Front suspension is by wishbones and coil springs, while at the rear is found the coil spring/damper "Chapman struts", with trailing radius arms. Apart from the immediate advantages of light weight and good roadholding derived from this system, there is a side advantage of relative simplicity for the amateur constructor; it must be borne in mind that the Elite is sold as a kit for home assembly as well as in complete form.

The ride is firm and exceptionally level, and there is only barely perceptible lean on corners, or "nose-diving" under heavy braking. Quite major irregularities in the road surfaces, taken at three-figure speeds, are marked only by a gentle rise and fall of the car on its springs, where other machines have occasionally left the ground with all four wheels; this movement is well damped, and there is no sign of pitching. Minor corrugations are dealt with as efficiently, and the only discomfort under these conditions was, as at low speed on rough surfaces, a fierce kicking through the direct steering as the front suspension does its work. The body/chassis unit remains rigid and inflexible under all conditions. Not directly associated with the suspension, but nevertheless a point which can be appropriately introduced at this point, is the vibration and shake of the body: at about 30 m.p.h. a low-frequency body shake gave the occupants a fairly uncomfortable time, while at higher speeds a high-frequency vibration set in which caused the entire car to reverberate. Allied to this, and probably the Elite's most serious shortcoming, is the amount of road noise which comes through to the occupants. This is "telephoned" up the rear suspension struts, from whence it is transmitted throughout the body shell so that, although wind noise is negligible, even at over 100 m.p.h., conversation is difficult at high speed—a factor which is not only extremely fatiguing after long periods, but which must be regarded as a major disadvantage in a "grand touring" car.

On the credit side, the road-holding is absolutely faultless on wet or dry surfaces. Light and extremely positive steering combines with an ability to hold a line through a corner with remarkable accuracy, enables the driver to place the car wherever he wants it; at the same time it is a forgiving machine, and over-exuberance, especially on wet roads, is not dealt with as severely as one might expect. The rear suspension in particular is extremely efficient, and wheelspin is virtually impossible to obtain.

Disc brakes are fitted to all four wheels, those at the rear being mounted inboard. Pedal pressure required is extremely high, but once this has been accepted it is realised that the stopping power is more than adequate for the car's performance. Only with difficulty can the wheels be locked, even on wet surfaces, which make a major contribution to the car's safety. A disappointing feature is the hand-brake, which is of pull-out type and, on the test car, barely effective on level surfaces and completely incapable of holding the car on any sort of gradient.

The interior of the car is smartly and functionally styled in black. Visibility is a little restricted by today's standards in that the waist-line is a trifle higher than is currently fashionable, and the windscreen, already shallow, is steeply raked and limits even the forward visibility of tall drivers. The instruments are laid out sensibly, however, lavish use is made of non-reflective surfaces, and the hand-controls are well-placed for easy identification and operation.

As might be expected of a car with the Lotus pedigree, the driving position is nearly ideal. The seats are comfortable, and provide excellent location for the shoulders and thighs, and the passenger is securely held, even without safety belts, against lateral forces under fast cornering conditions. The small-diameter steering wheel is well-placed in terms of height and plane, and although endowed with disconcertingly little castor-return action, the steering is light, precise and suitably high-geared. The driver has plenty of elbow room, and the gear lever is well-placed, on top of the transmission tunnel, for maximum use; one has only to sit in the car to wish immediately to take it out on the winding, climbing and falling roads on which the car is so very much

Continued on page 79

Cars on Test

LOTUS ELITE

Engine: Four-cylinder, 76.2 mm. × 66.6 mm. (1,216 c.c.); compression ratio 10.0 : 1; single overhead camshaft; two S.U. H.4 carburetters; 85 b.h.p. at 6,300 r.p.m.
Transmission: Single dry-plate clutch; four-speed and reverse gearbox with synchromesh on all four forward speeds. Central remote control gear lever.
Suspension: Front, independent by wishbones with coil springs and anti-roll bar. Rear, independent by Chapman strut and wishbone system. Tyres: 155 × 15.
Brakes: Front, Girling hydraulic disc, outboard; rear, Girling disc, inboard.
Dimensions: Overall length, 12 ft. 4 ins.; overall width, 4 ft. 11¼ ins.; overall height, 3 ft. 10½ ins.; turning circle, 29 ft. 6 ins.; ground clearance, 4¼ ins.; kerb weight, 13¼ cwt.

PERFORMANCE

	m.p.h.			secs.
MAXIMUM SPEED	—117.1	ACCELERATION	0–30 —	4.0
			0–40 —	6.3
(mean of 2 ways)	—116.1		0–50 —	8.2
			0–60 —	11.0
			0–70 —	15.0
SPEEDS IN GEARS First	— 47.0		0–80 —	19.5
			0–90 —	23.8
Second	— 66.0		0–100 —	33.0
Third	— 90.0	Standing quarter-mile		—18.25

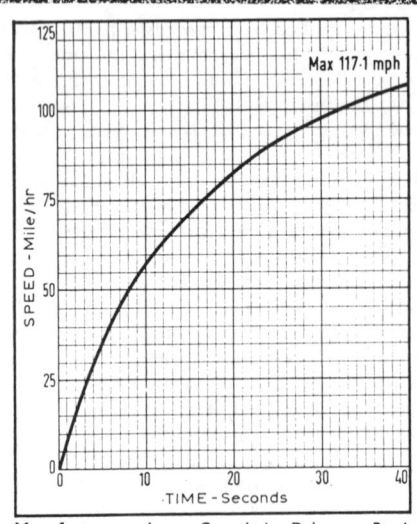

Max 117.1 mph

Manufacturers: Lotus Cars Ltd., Delamare Road, Cheshunt, Herts.
Price: (As kit of parts), £1,375; as complete car, £1,662 0s. 2d., including purchase tax.

LOTUS ELITE CONTINUED FROM PAGE 78

at home. Set immediately in front of the driver are the large 140 m.p.h. speedometer and matching 8,000 r.p.m. tachometer while fuel level, oil pressure and water temperature gauges are disposed beneath and at the sides of them. A combined headlamp flasher/horn switch is fitted, and the instruments are provided with variable lighting. Wet weather visibility is rather weakened by two-speed windscreen wipers which, disappointingly, lift off the screen at high speed.

Any doubts one may have about the longevity of the Elite design can be immediately settled by an examination of the test car. When it came into our hands it had covered some ten thousand very hard miles; our test added a further thousand to this total, and the car was handed back to Lotus without any squeaks or rattles. No cracks were visible in the glass-fibre, and the body and doors were draught-free.

The performance is, naturally, one of the Elite's most attractive features. A maximum speed, taken as the mean of runs in opposite directions, endow the car with a genuine three-figure cruising speed, although at this velocity the noise level is uncomfortably high, the car is at its best from the passenger's point of view at around 90 m.p.h. At this speed the mechanical components are working well within their capacity. Acceleration, of course, is splendid, and from rest it takes very little more than half-a-minute to reach 100 m.p.h. The well-conceived shape permits the maintenance of extremely high average speeds at modest fuel consumption, and an overall figure for our test mileage of 30 m.p.g. is almost startling when it is considered that full use of the performance was made whenever possible. At 6½ gallons capacity, the fuel tank is barely large enough for real long-distance fast touring, but it is not embarrassingly small.

Taken all round, the Elite is a most satisfying car to drive. Such a small, compact car, with light weight, generous power and superb roadholding sets new horizons for G.T. car designers.

LOTUS ELITE

An Appreciation and an Obituary

AFTER SIX YEARS of production, the Lotus Elite is no more. Gone it is, the way of the Allard J-2, the L-29 Cord, the Mercedes SS-K and other designs of sacred memory.

A beautiful design was the Elite, one of the great designs of the post WW-II era, one that seems certain to be looked back upon as a landmark of some sort in automobile design. Without question, it was one of the best, if not the very best looking Grand Touring car ever built. The body, all fiberglass, was designed entirely in the Lotus works at Luton, England, and was not only an immediate and lasting success but also an example, perhaps the only example, of the fluid plasticity of speed/motion to be captured in that glass fiber and resin medium.

Admittedly, the Elite had its problems. The monocoque-type construction created a drumming that was extremely annoying to most passengers, the body had minimal (if that) protection from the expected hazards of normal driving, and the average mechanic went into shock when asked to work on it.

The car was also plagued by other problems—a high initial price that scared off all but the most sanguine, a marketing situation during its introduction that could only be described as impossible, and a long period (before Bob Challman of Ecurie Shirlee Corp. stepped in to give the American Lotus plan a respectability and dependability it never had before) when the buyer of an Elite didn't know whether he'd ever see his dealer again.

So, think well of the Elite. It will be remembered as one of the outstanding designs.

Charge of the

50 YEARS

With the launch of the Seven and Elite, Lotus entered the fray as a serious sports car manufacturer. **Martin Buckley** discovers that these small cars of the Fifties still offer refreshing purity on today's roads

light brigade

THE SEVEN AND THE ELITE, BOTH LAUNCHED IN 1957, ARE PERHAPS THE most critical machines in the history of Lotus as a serious builder of road cars. As a blend of scintillating dynamics and aching beauty, one of them, at least in my book, remains unsurpassed by any Lotus since. No longer willing to rely on the seasonal vagaries of racing car construction, Colin Chapman saw his future – and fortune – as a year-round builder of roadgoing sports cars, bringing his ethos of minimalism and lightness to cars you could use every day.

The Elite, advanced and beautiful, was to be his new GT car, cast in the elegant mould of upmarket foreigners like Porsche and Alfa. The sketchy Seven, at the other end of the scale, was pitched as a cheerful build-it-yourself device that bobble-hat-wearing enthusiasts could race on Sunday and drive to work on Monday.

The bare necessities of life in a Lotus Seven...

Recognise the Austin A35 sidelights?

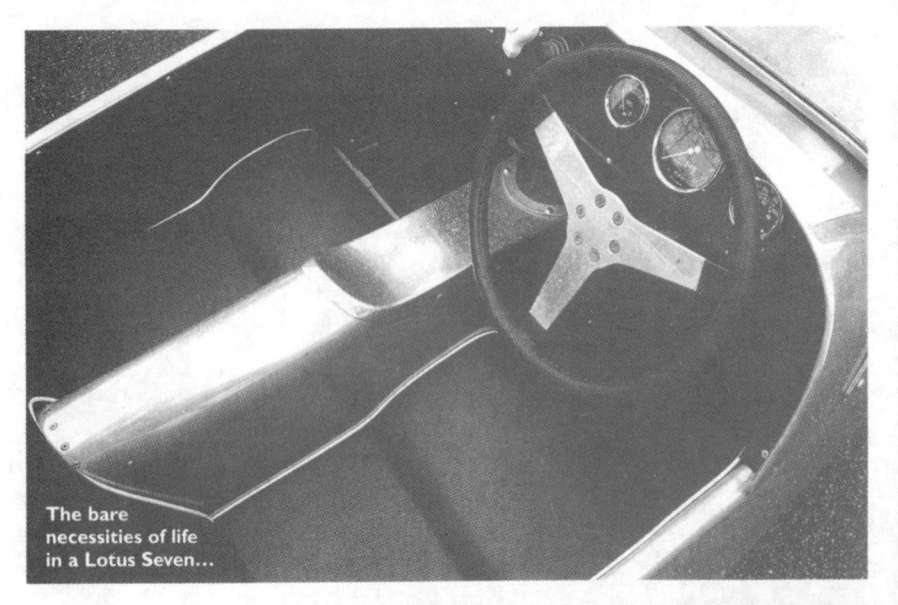

Even on 55bhp, early Seven – weighing in at a mere 960lb – is a gas

In the event, the success of the latter bankrolled the failure of the former: the Elite was a classic case of a car's specification being beyond the ability of contemporary technology to build it reliably. Its alarming growing pains – including broken diff carriers and stress cracks around the pedal box – are legendary.

Forty years on, there is a refreshing tang of purity about these cars and the way they go. They fit like gloves, bereft of the flab and excess that make so many other Fifties sports cars such a misery to drive.

Our Seven is a late Series 1 – a Cheshunt rather than a Hornsey car – from the personal collection of Caterham boss Graham Nearn. Bare polished aluminium adds to the functional appeal of its spare shape, characterised on the Series 1 by simplistic cycle wings enfolding slender tyres, allowing a good view – even from the driver's seat – of the twin-wishbone front suspension and the little cast-iron Ford brake drums.

At £536 for the kit, this was raw driving appeal at its most unadorned, with little to distract the attention: no hood, no side screens, no heater, feeble wipers and only the basic information – amps, oil pressure, temperature and speed – from the dash. Daily thrills for the failed racing driver with a strong streak of masochism.

Hop over the side and slide down on to a basic, shapeless, red plastic cushion with a one-piece backrest. Hips 'twixt propshaft tunnel and outer body, right elbow hanging over the side, you feel snug with everything you need within reach, but vulnerable to the ravages of wind and road. The steering column – leading to an upside-down Morris Minor rack – is very tight against the clutch: I had to drive the Seven in stocking feet, so tiny and closely spaced are the pedals.

Most early Sevens had Ford sidevalve power but this one has a mildly tweaked twin SU-fed Morris Minor A-Series, good for about 55bhp and barking aggressively out of a side exhaust that blackens the aluminium of the rear mudguard with its breath. Deploying its power through a nice-to-use four-speed BMC 'box, it will pull from low speeds in top, although a combination of sharp

PHOTOGRAPHY BY TIM ANDREW

Watersports: you 'ad to be 'ard to drive a Seven in winter

'Daily thrills for the failed racing driver with a strong streak of masochism'

clutch bite and initially hair-trigger throttle action can initiate a series of kangaroo-hops when you want to trickle in traffic.

Hauling the 960lb Seven, the little 948cc four runs out of breath at 80mph but gets there in short order, that sharp throttle response adding to the impression of eager sprinting ability.

You drive the Seven with toes and fingertips rather than hands and feet, your reactions as second nature as blinking or taking breath. Caress the smooth, light, instant steering and feel the clean bite of the front wheels as you turn in, savouring the Seven's relish for every kind of corner – except perhaps the bumpy ones, which make the tubular frame whip as the beam rear axle becomes unruly.

If the Seven is still great fun, then the Elite's virtues run deeper as a proper all-rounder offering long-distance space and comfort for two.

Its aesthetic near-perfection, so balanced and delicately drawn, is all the more amazing when you consider that it is the work of novice stylist, turned chartered accountant, Peter Kirwan-Taylor. What's more, with the help of aerodynamicist Frank Costin, the flat-bellied Elite has an amazingly low 0.29 drag factor which explains why, on just 72bhp in basic form, it managed over 110mph and 40mpg.

It is as easy to appreciate the Elite with your head as it is with your heart. Chapman's notion of an ultra-lightweight glassfibre monocoque –

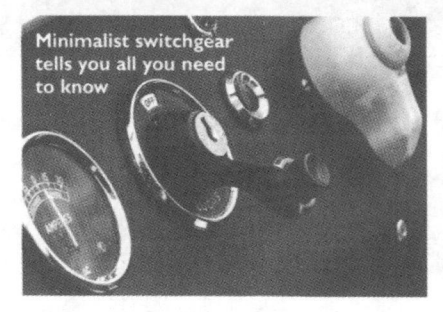

Minimalist switchgear tells you all you need to know

double-skinned and box-sectioned for strength, and weighing under 1500lb when complete – was ambitious but wonderfully brave, especially when you remember that this was his first experience of this newfangled material.

That the car handled and rode so brilliantly would have come as no surprise to anyone who had driven a contemporary Lotus racer: the Elite's elegantly simple Chapman strut rear suspension, isolated from the chassis by rubber, with the loads spread as wide as possible to prevent fracturing, was a direct lift from the Type 12 Formula 2 car, as were the front wishbones.

Geoff Thompkins fell in love with the Elite when a friend was given one as a present in 1961. 'He had it for about a year and had great fun sprinting it,' recalls Geoff, who didn't think about owning one himself until 25 years later, casting around for a classic to restore in his spare time.

His concours-winning Special Equipment car –

A-Series from Morris Minor gives extra 15bhp as standard

with Bristol Plastics-built body rather than the earlier, cruder, Maximar-fabricated effort – has now been on the road since 1989, after a total rebuild that included a complete restoration of the shell by Elite specialist Tony Bates.

The simple elegance of the exterior is reflected in the Elite's cabin, styled by moonlighting Ford designer Peter Cambridge (who apparently never got paid by Chapman). Whether the elongated silver panel that houses the instrumentation ⇨

Interior of Elite is thoughtfully worked out

Dash signed by interior stylist Peter Cambridge

was intentionally styled to match the side profile of the Elite or not, it is still a nice touch, set off by a lovely three-spoke wood-rimmed steering wheel with a ridged circumference that is a cool joy to caress. Cambridge's high-backed seats are comfortable despite thin padding, while his hollowed-out door pockets that liberate elbow room, and the crafty lift-out windows that can be stored in the backs of the seats when you want some ventilation are utterly rational.

Legroom in this semi-recumbent driving position is ample, and unlike the Seven there is lots of room in the footwell for my size tens to dance on the hung pedals. Behind you, either side of the spare wheel, the top mountings for the rear suspension struts jut out aggressively like Madonna's brassiere.

The Climax FWE engine, son of a lightweight firepump unit designed to run at maximum revs from cold, gives 83bhp on a pair of twin-choke Webers. It fires with an eager and aggressive bark and it is soon clear that there is little in the way of sound-deadening between us and it. Lesser Elites had the MG Magnette gearbox but this SE has the ZF unit specially built for the model, and it needs to be used liberally to make use of the engine's modest low-down torque. Luckily it's a gem of a 'box, its flick-switch action requiring fingertip effort for precision movement in a short-travel gate.

The clutch is short in its travel too, and a little sharp, so smooth take-offs require practice. Despite this lack of torque, the engine gets into its stride as the revs rise, with no apparent step in its delivery.

Keep the revs up, then, and the Elite performs strongly: it should get to 60mph in a respectable ten seconds and touch 112mph in this tune. Judging by the racket it makes at 60, you'd need earplugs at three-figure speeds, though Geoff reckons you get used to that even on a long run. Not that it is an unpleasant noise, the bark of gobbling Webers mixing with the shrill whine of timing gears for a hard-edged war cry that is pure racer. In addition, the nature of the construction allowed for little isolation from road and suspension noises, and you can hear the struts working through

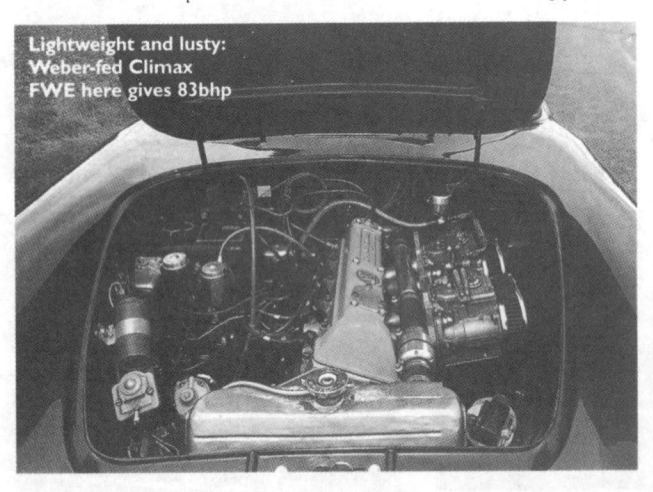

Lightweight and lusty: Weber-fed Climax FWE here gives 83bhp

Godiva was trademark of Coventry Climax

Beautiful and slippery: drag factor of 0.29Cd not beaten for 20 years

"Pre-motorways, there were few quicker ways to travel than in an Elite"

Elite's poise unsurpassed 40 years ago: still handles great today

	SEVEN	ELITE
Engine	948cc	1216cc
	4cyl OHV	4cyl SOHC
Power	55bhp @	83bhp @
	4000rpm	6250rpm
Torque	50lb ft @	75lb ft @
	2500rpm	4750rpm
Weight	960lb	1455lb
0-60mph	12.1sec	10sec
Top speed	83mph	112mph
Value	£13,000	£25,000

Side windows removable for ventilation

those mountings in the back – but who ever bought a Lotus Elite for its refinement?

It's the handling that makes the car, with steering that is sharp and informative, accurate enough to make the best of the Elite's wonderful neutrality and poise along sinewy rural roads.

Set at your ease by the driving position that leaves none of the essentials too far or too near to inefficiently manipulate by hand or foot, the Elite fits like a favourite slipper. The slender rubber gives all the adhesion you need and, as you begin to work into its rhythms, and with so little weight to anaesthetise and smother its superlative feel, the tiny Lotus begins to feel almost organically synthesised with your responses. The ride has a wonderfully light touch over all kinds of surfaces, a supple combination of stiff damping and soft springs that gave a level of comfort that some big, heavy saloons of the day struggled to match, never mind sports cars.

It's not difficult to see why so many owners put up with the Elite's problems and were willing to pay such a high price for the car – more than an E-type Jaguar, although from 1961 the availability of the car as a kit cut the price dramatically. Two years later, with the much less radical Elan on the scene, Lotus dropped the Elite with a hard-to-stifle sigh of relief: it wasn't a money-maker for Chapman or the companies he had commissioned to build the labour-intensive bodywork.

Nobody knows for sure, but about 1000 Elites eventually took to the road, if you take into account the 30 or 40 shells Lotus had left over when production stopped. Around 600 survive around the world, with specialists like Miles Wilkins and Tony Bates dedicated to keeping this delightful but difficult-to-live-with car in regular use. From a low point in the late Sixties and Seventies – when £300 would have bought a runner – really good Elites were drawing amazing figures in the late Eighties (Geoff Thompkins was offered £45,000 by a Japanese Lotus freak), but have currently stabilised at around £25,000 for a prime example. Deservedly, the Elite is the most valuable of the roadgoing Lotuses.

The Seven's fate is well documented. As the brave but deeply flawed Elite began to drop out of sight, the Seven carried serenely on, with timeless appeal as a pure and unadorned fun car with few compromises, although it has taken the dedication of Caterham to raise it to cult status.

If you like this sort of car, nothing else will do – but it's the Elite that is the only Lotus I've ever hankered after. To have owned one in the late Fifties would have put you on a dynamic plane way above mortal motorists, enjoying steering, grip and control previously found only in pure-bred racers. Jaguars and Astons may have been faster in a straight-line burn-up, but nothing could touch an Elite on twisty rurals. Pre-motorways, there were few quicker ways to travel. ●

CLASSIC PROFILE
FEATHERWEIGHT

CHAMPION

With its lightweight all-plastic body and all-alloy Coventry Climax engine, the Lotus Elite was revolutionary at its launch in 1957 – and an instant winner. Mark Hughes identifies its strengths and weaknesses; pictures by Andrew Morland

If the Lotus Elite was being announced now, in 1988, it would send shock waves through the motor industry. Considering that its specification and capability bear comparison with the world's best today, imagine the impact it had on the motoring public back in 1957 on its launch at the Earl's Court Motor Show.

Its novel glass-fibre monocoque construction was unique then, and remains so today. No other car has ever used glass-fibre *throughout* for its strength. Even the engine, gearbox and final drive are bolted directly to glass-fibre, only the front suspension picking up from a small steel subframe bonded into the structure. The durability of Elite bodyshells now nearing 30 years old bears witness to the qualities of a mode of construction which makes the conventional steel monocoque seem antiquated.

Aerodynamics are in vogue these days, yet only a tiny handful of modern cars – Audi 80, Vauxhall Carlton and Renault 21 – can match the Elite's drag co-efficient of 0.29. Not only does the Elite's delicate profile slip through the air like a paper dart, but its flat belly ensures that underbody aerodynamics work well too.

Great strides have been made in fuel economy in recent years, yet everyone forgets that Colin Chapman produced a 115mph car which can exceed 40mpg with ease. In their higher states of tune Elites can cover 0-60mph in less than 10sec, yet engine capacity was only 1.2 litres. All the modern cars which can achieve this level of performance from such a small capacity need a turbocharger to help them along.

Current motoring magazines are full of the advances Citroën has made in lightweight construction with its flimsy new AX, yet that car's 1450lb weight is equalled by the Elite. Consider, too, how advanced for 1957 was all-round disc braking and all-independent suspension. Whatever yardstick you select, the Elite was born years ahead of its time…

It was the Porsche 356s and Alfa Romeo Giuliettas of the time which Chapman had in his sights, both on the tracks and in the marketplace, when he conceived the Elite in 1956. Success in the FIA's new 1300cc GT class, at Le Mans in particular, would be the promotional launchpad to take Lotus into this higher league, and away from a reliance on racing car manufacture.

Since a steel-shelled car would have been too heavy for his power-to-weight requirements, and the tooling-up was beyond Lotus's resources, Chapman toyed with the idea of using glass-fibre, a technology which Jensen, Chevrolet and Turner had recently proved. But he wanted to go one step further and make the *entire* bodyshell from glass-fibre. It was an ambitious target, but Chapman had enough self-confidence – and the application to teach himself by making small-scale experimental structures – to carry it off.

One of his friends was a 27-year-old accountant called Peter Kirwan-Taylor, who as a boy used to sketch car profiles in school exercise books. Although not a car stylist by training, he nevertheless had produced a handsome body for his own Lotus Mk6. Chapman was impressed, and invited Kirwan-Taylor's ideas for the new car. Years later, as a high flier in finance, Kirwan-Taylor would admit to being influenced by Pinin Farina's Ferrari Superfast and

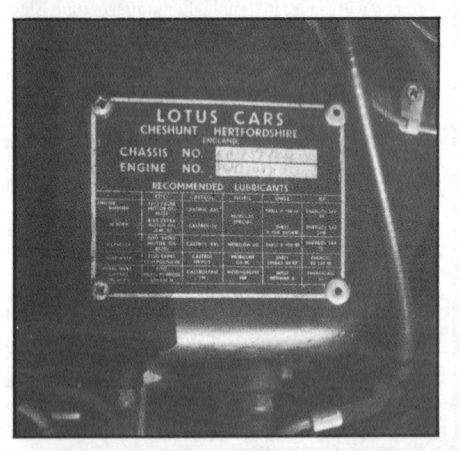

The interior of the Elite (top left) is elegantly simple, and entry and exit are helped by wide-opening doors (top right)! Engine is the classic all-alloy single ohc Coventry Climax FWE (Feather Weight E) unit, pushing out between 75 and 90bhp, depending on tune, enough to give the sleek little coupé a sparkling performance and to make it a demon of the race tracks – Jim Clark among others raced one successfully. Part of the reason for the excellent performance was the beautiful and slippery (Cd = 0.29) body, designed by Peter Kirwan-Taylor (below)

Bertone's BAT Alfa, but functional design, without the fins and chrome which were all the rage, was his aim.

Kirwan-Taylor's shape was not altered substantially as the Elite edged towards reality, although Frank Costin was called in to refine its aerodynamics, most notably by suggesting a Kamm tail. John Frayling, a sculptor from New Zealand who had worked in Ford's design department, then set to the job of producing first a one-fifth scale clay model, then a full-sized plaster one, in tiny overspill premises in Edmonton, just down the road from Lotus's Hornsey works. According to an apocryphal story, space was so tight at Edmonton that this plaster model was eventually thrown out by pushing it into a nearby railway wagon – it could still be circulating somewhere on Network South-East!

The engine Chapman selected for the Elite was Coventry Climax's all-alloy, four-cylinder, single overhead cam unit which Lotus had used with great success in 1460cc FWA guise in its sports racing cars. With its origins as a fire pump developed for use in the Korean War, this was a free-revving engine designed with the wide tolerances necessary to enable it to run at high revs straight from cold. It sounds a crude ancestry, but its pedigree was unquestionable, with Harry Mundy and Walter Hassan responsible for its design – they would later produce the Jaguar V12.

For the new Elite, Coventry Climax developed a special FWE (Feather Weight Elite) road-going version with more tractability by amalgamating parts from the FWA sports racing and FWB Formula 2 engines. So it was that the FWE took the FWA's 66.7mm stroke and the FWB's 76.2mm bore to arrive at a 1216cc unit, which was strong on both torque and power. Cast iron was considered for the block to reduce production costs, but in the end rejected because of the investment needed. As well as the block and cylinder head, the FWE's sump, fan pulley, water pump and timing and cam covers were made from aluminium. With a single SU carburettor, this engine developed 75bhp at 6100rpm, yet weighed only 215lb dry.

Taking the Elite from prototype to full scale production occupied the best part of 18 months

Chapman used his racing knowledge in developing the Elite's suspension. There was a simple double wishbone layout with Armstrong coil spring/damper units at the front, while at the rear there were struts evolved from the F2 Lotus 12. These Chapman struts protruded high into the cabin and carried the spring/damper units, the driveshafts providing lateral location and a kinked radius arm mounted to the hub carrier the fore/aft location. It was a simple, light and effective layout, even if the Chapman struts did consume cabin space. Braking was thoroughly avant-garde with discs all-round, those at the back mounted adjacent to the final drive.

Peter Cambridge designed the Elite's functional but comfortable interior, taking a shape similar to the car's profile for the instrument binnacle, which contained two large dials for speed and revs, and three smaller ones for fuel, water temperature/oil pressure and battery charge. There was no space for rear seats, but the platform above the spare wheel well behind the front seats usefully supplemented the sizeable boot. Although Chapman was obsessed with using fixed side windows, his design staff persuaded him to accept removable plexiglass windows which could be stowed in pockets on the back of the front seats. The car would demonstrate its superb aerodynamics by allowing no more than a gentle breeze through open windows at 100mph.

Just two months before the new Elite was due at Earl's Court, moulds were taken off the completed plaster model and the first glass-fibre prototype laid up in four sections. After these had been bonded together like a giant plastic kit, the interior panels and bulkheads were fitted in 56 small sections. Another Ford refugee, Ron Hickman, became involved in handling production engineering, while a certain budding racing driver named Graham Hill was the foreman at Edmonton.

Like all things Lotus, the final stages were a mad rush. Two days before the Earl's Court press preview, this first shell hadn't even been trimmed or fully painted, yet mere hours before the show opened the cobbled-up prototype, finished in silver with a metallic grey roof, was trailered to Alexandra Palace for some quick publicity photographs. Undoubtedly the Elite was the star of the show, but even now there was a problem. The single key wouldn't unlock either of the doors, which was no great difficulty, until Princess Margaret and Lord Snowdon asked to sit in the car. Lord Snowdon and Kirwan-Taylor reportedly spent many hours trying to burgle their way in…

Taking the Elite from prototype form to full-scale production occupied the best part of 18 months, as Lotus struggled to make the glass-fibre monocoque practical for quantity production. Eventually a boat-building firm called Maximar was hired to build bodyshells, and Lotus engineers, including Frayling, worked at Maximar's Sussex base to finalise production design. There were always quality problems, and Maximar-built shells can vary in length by ¾in and in width by ½in…

Although that original show car was later broken up after a period on display at Lotus's Hornsey showroom, the second prototype was delivered to a racing customer, Ian Walker, in May 1958. The day he took delivery, he drove up to Silverstone, entered the 1600c class of the *Autosport* production sports car event, and won…

This was the first win in the Elite's magnificent career of competition, and the cars won their class in all six appearances at Le Mans between 1959-64. Peter Lumsden/Peter Riley finished eighth overall in the first year, and two more Index of Thermal Efficiency awards came in 1960 and 1962. Frank Costin developed even better aerodynamics for these race cars, and the highest speed clocked on the Mulsanne Straight was a staggering 138.2mph… from just 1300cc. Elites were raced at home too, most successfully by the famous duo of Les Leston with DAD10 and Graham Warner in LOV1.

Lotus records show that on December 31, 1958, the first two production Elites were sold, to band leader Chris Barber and Ian Scott-Watson (Jim Clark's mentor) of Border Reivers, but manufacture was slow to get under way in the cramped Edmonton shop. It was not until Lotus's purpose-built new factory at Cheshunt opened in October 1959 that Elite production began in earnest, two full years after its launch.

Above: The original publicity photograph, taken hours after the car was finished, and just prior to delivery to the 1957 Earls Court show. Left: All ship shape and Bristol fashion – model, Elite and Bristol freighter

Despite the high price, the Elite's labour-intensive bodyshell manufacture meant that Lotus lost money on every one

With Coventry Climax supplying complete engines which were then mated to BMC's B-series four-speed gearbox, and Maximar sending painted and trimmed bodyshells up from Sussex, Cheshunt was really an assembly operation producing anything between eight and 36 cars a month. At last the Elite was in volume production, costing £1949. It was undeniably expensive – a Jaguar XK150 or E-type always cost much the same – but was well received by the motoring press, which heaped praise upon its technical sophistication, brilliant handling, lively performance, gorgeous styling and superb brakes.

Despite the high price, the Elite's labour-intensive bodyshell manufacture and the expense of Coventry Climax engines meant that Lotus lost money on every car it built, which explains why the Elan evolved into such a different, cheaper car. Chapman, however, did rather well, the Elite putting him well on the way to amassing his personal fortune…

The most significant specification changes came in 1960 with the Series 2 Elite. As a result of the deal Chapman had negotiated, Maximar was finding the build of its contracted 250 bodyshells financially crippling, and quality was suffering. Chapman's search for a new supplier ended with the Bristol

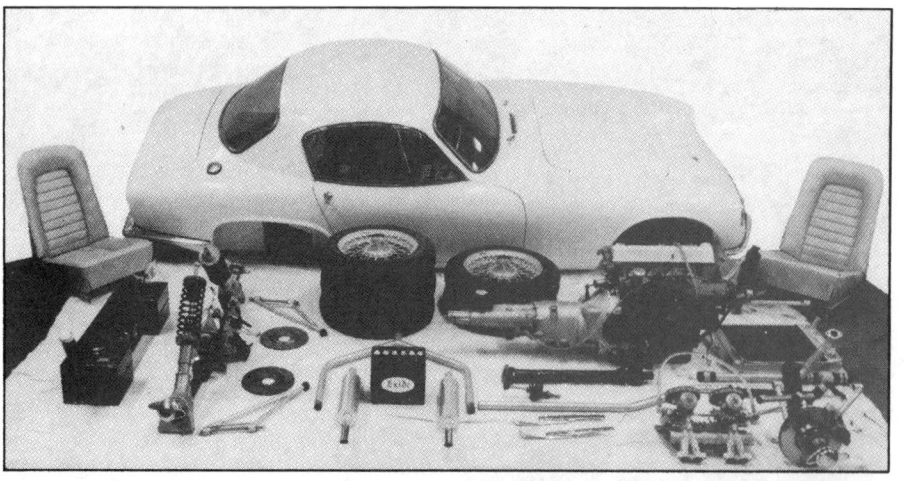

Left: Lotus Elite in component form with 83bhp twin SU specification engine. Above: Compact boot, with side-mounted battery, of the original Earl's Court show car

Aeroplane Company, which was pleased to accept an association with a now prestigious automobile name. Bristol, too, found the Elite more than it had bargained for, and it healthily subsidised Lotus through the early sixties! Quality was never allowed to suffer, and Bristol bodyshells (identified by a plate under the bonnet) were always beautifully made.

This change also presented an opportunity to make other modifications to the Elite. The crooked radius arm rear suspension location had always induced a rear wheel steering effect during hard cornering, so for the S2 a triangulated wishbone was substituted. The interior was also tidied up, an attractive semi-rigid thermoplastic called Royalite now being used for door and transmission tunnel trim.

Shortly after the S2 came a Special Equipment (SE) version for buyers who wanted more performance. Twin SU carburettors and a fabricated four-brach exhaust manifold boosted power to 83bhp at 6300rpm, but more was to come. The Elite SE could be specified with a Stage II Climax producing 90bhp by using a higher-lift (but still three-bearing) camshaft, ignition alterations and an 11.0:1 compression ratio achieved by shaving 40 thou off the head. Capitalising on racing experience, Lotus also built short runs of even faster Elites: 23 Super 95s (the figure indicated horsepower) were made with twin Weber DCOEs instead of SUs, and there were six each of the Super 100 and Super 105. These ultimate derivatives had fully balanced engines, a high-lift five-bearing camshaft (which could rev to 7900rpm without whipping) and an 11.5:1 compression ratio.

All standard Elites came with four-speed BMC gearboxes using the same ratios as the MGA. Although a rugged unit, it has no synchromesh on first gear, and its terrible step between second and third makes it unrewarding in fast driving. With the S2, therefore, SE models were offered with a close-ratio ZF four-speed 'box. With a light and precise change, and better ratios, it's much pleasanter than the MG 'box, although the combination of the engine's lack of low-down torque and the ZF's tall first gear makes hill starts tricky.

Despite its remarkable qualities, the Elite was always a hard car to sell, on account of both its price and the lack of a good dealer network. Lotus employees remember the fields behind the Cheshunt factory chock-a-block with bodyshells waiting to be built up once there were a few orders. Pre-Elite, Lotus had always sold cars direct from the factory, and its initial stumbling block was a lack of a dealers. Good ones, like David Buxton and the Chequered Flag, were appointed in England, but this was a much tougher nut to crack in the USA.

After a difficult period when Jay Chamberlain handled Lotus imports in the USA, Chapman managed to forge a deal with existing BMC distributors, but only by the time the Lotus name had been tarnished and the Elite's days were over. With no dealers to turn to, the American Elite owner was sunk when his car went wrong. This depressed sales to such an extent that cars could sit around for as long as two years before they found buyers...

The Elite's price made sales sluggish in England too, but by offering the car in component form in 1961 Chapman was able to shift more Elites. By this time, however, Ron Hickman's design for the new Elan was well on the way, and lessons had been learned from the Elite about manufacturing costs. It is commonly thought that the glass-fibre monocoque was discarded for the Elan because of inherent inadequacies, but this is far from the truth.

From the beginning, the Elan was intended also to have a glass-fibre monocoque. Its steel chassis was tested in prototype form only as a development convenience, to provide mounting points for running gear while a glass-fibre test shell was engineered. The steel backbone worked so well, and was so relatively cheap to make, that the Elan project about-turned in this new direction...

The Elite's reputation went through a grim period in the late sixties and seventies as the owners of well-used cars, invariably on low budgets, struggled with the Coventry Climax's pernickety nature. Often they neglected the obsessive maintenance this highly-strung engine requires, forgetting that it consumes oil almost as quickly as other cars drink petrol, that regular tightening of cylinder head, differential and suspension nuts and bolts is necessary, and that universal joints need frequent greasing.

Thankfully, most Elites have now moved on from gorilla owners and into the hands of enthusiasts who understand their demanding nature. The Elite has emerged from the reputation depths and now soars high, commonly recognised for the design landmark it is.

BUYER'S SPOT CHECK

Just because the Elite preceded the Elan doesn't mean that there is any similarity between the cars. Indeed, the Lotus badge is about the only link, for the Elite's body structure, engine, gearbox, final drive and brakes are completely different. Whereas much of the restoration of an Elan is within the capability of a competent do-it-yourself owner, the Elite is a much more formidable task which calls for a professional level of expertise. As a result, the process of selecting your car in the first place requires thorough study.

You should start with a very careful examination of the body, as bodyshell restoration is an expensive job requiring great skill and patience. You can reduce labour costs by carrying out pre-painting surface preparation, but don't consider glass-fibre patching of serious damage, as the car's structural integrity depends on expert attention. It's wise, too, to

SPECIFICATION	LOTUS ELITE
Engine	In-line four
Construction	Alloy block and head
Bore/stroke	76.2mm × 66.7mm
Capacity	1216cc
Valves	Single ohc
Compression ratio	10.0:1
Fuel system	Single 1.5in SU, twin 1.5in SU (SE) or twin Weber carburettors (Super 95)
Power	75bhp at 6100rpm (single SU) 83bhp at 6300rpm (twin SU)
Torque	77lb ft at 3750rpm (single SU) 75lb ft at 4750rpm (twin SU)
Transmission	Four-speed manual (MG or ZF)
Brakes	Girling 9½in discs all round, inboard at rear
Suspension front	Ind by double wishbones, coil springs, Armstrong telescopic dampers, anti-roll bar
Suspension rear	Ind by Chapman strut, fixed length driveshafts, trailing arms (triangulated wishbones on S2), coil springs, Armstrong telescopic dampers
Steering	Rack and pinion
Wheels/tyres	Knock-on 48 or 60 spoke wire wheels, 4.80-15 tyres

DIMENSIONS	
Length	12ft 4in
Width	4ft 11¼in
Height	3ft 10½in
Wheelbase	7ft 4½in
Unladen weight	1456lb

PERFORMANCE	
Maximum speed	112mph
0-60mph	11.4sec
Standing ¼-mile	18.4sec
Fuel consumption	34mpg

consult with the paint shop you plan to use, as they may want to carry out the final surface preparation.

Removing existing paint involves several applications of a water soluble stripper, and much laborious scraping by hand to reach the gel-coat. Then gel-cracks have to be ground out and the required glass-fibre thickness restored with fine matting. The exact contours are achieved by applying polyester filler, which then must be rubbed down with wet and dry paper to provide a key for the new paint. Opinions are divided about whether it is best to use cellulose paint or a modern acrylic two-pack finish.

Like any other glass-fibre car, all but the most thoroughly restored Elites will have suffered the ravages of heat, water and frost over the years, causing delamination and gel-cracking. You must assess the deterioration, and bear in mind the cost of a 'respray' – at least £4000 if a professional does all the work. Bumps and dimples indicate short-cut filling and repainting where the gel has cracked, and a proper strip-back and respray will eventually be necessary. Any accident damage always needs expert remedial work, so do look carefully for amateur patching. 'Weetabix' glass-fibre surfaces anywhere except on the bulkhead inside the car are evidence of bodged work, which is both dangerous and undesirable. As the Elite has a double-skinned glass-fibre structure, all sub-surfaces – underside, engine bay, boot and wheelarches – should have a smooth finish.

Just as much attention needs to be paid to the engine, as a rebuild is very expensive – as much as £2500 for a complete overhaul. Tony Mantle of Climax Engine Services can supply all engine items except blocks and cylinder heads (although he expects to have new ones available by the end of 1988), but the cost of parts is high: a few examples are a set of pistons at £240, liners at £120, a crankshaft at £550 and a three-bearing Stage II camshaft for £200. The company does supply racing Elites (invariably overseas) with completely new engines, and these can cost as much as £7000. Fortunately, for limited road use you should be able to preserve many parts from the most tired of engines. Thanks to Climax Engine Services' sole distributorship, the spares position is better than it's ever been.

If you've never driven an Elite before, don't be alarmed by its noise and vibration. An engine that sounds as if it's trying to shake itself out of the car is normal, and don't expect much pulling power below 3500rpm. Any service and rebuild records will be good evidence of the owner's enthusiasm for maintenance, and this is the best guide you have.

As it is all too common for a casual owner to forget to keep the engine well supplied with oil – consumption is heavy, at 150-300 miles per pint – check the sump level. You should look in the radiator, for oil can end up in the water if the engine has been allowed to overheat excessively through water shortage. Despite the engine's appetite for oil, there should be no blueness from the exhaust. Smoke on lift-off suggests worn valve guides, and on acceleration it means that the piston rings are tired. Tickover should be steady at 900rpm, and oil pressure should read 10-25psi at tickover, 40-55psi on the move. Make sure there are no water leaks when the car's warming up, and check that running temperature is around 65 degrees – these engines run very cool.

The expense of rebuilding a Climax engine, as well as the fact that a mere 35,000 miles is a typical distance between overhauls, means that skimping on maintenance used to be common with Elites, although their rarity and desirability has reduced this in recent years. You are unlikely to find trouble with an obviously pampered car, but anything less may show symptoms of ham-fisted attention. Unlike Lotus's twin-cam engine, the Climax is held together with metaphorical 'finger tightness', and it is quite common to find pulled threads where gorilla mechanics have over-tightened bolts without a torque wrench. The head is held down at only 20lb ft,
the main caps retaining the crankshaft need 38lb ft, and the big ends 22lb ft. Horrifying though it sounds, a cracked head is quite common – around the dynamo mounting is the crucial place to look for signs of anti-freeze staining and water seepage.

Transmission faults are the least of your worries with an Elite, unless you are examining an S2 car with the optional – and more desirable, on account of its slick gearchange – ZF close-ratio four-speed gearbox. Many ZF parts are no longer available, but this gearbox is so understressed in the Elite (it could easily withstand 2½-litre Climax power in the Lotus 15) that serious problems are unusual, unless an owner has been running without oil or tried to change gear without the clutch. Only seals and bearings are generally needed. Parts for the more common, lorry-like MGA 'box are plentiful, and complete gearboxes can often be found with owners who have fitted a ZF replacement.

Listen for the usual symptoms of gear whine, clonking from the differential and worn synchromesh, but use these as bargaining points rather than reasons to reject a car. Contrary to the many myths that surround the Elite, differentials do not pull out of the glass-fibre, which is ¾in thick in this area. One pre-production car had its differential rip clear of the bodyshell, but no-one knows of this having happened since. If there is a noise from the final drive, check

Below, top to bottom: 'Super 95' engine with twin Weber set up. Check steel subframe and suspension mountings for corrosion as replacement is very expensive. The original prototype Elite bodyshell at Edmonton before fitting out

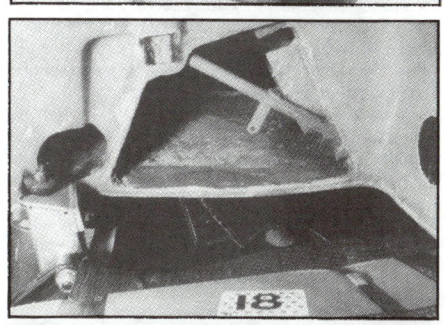

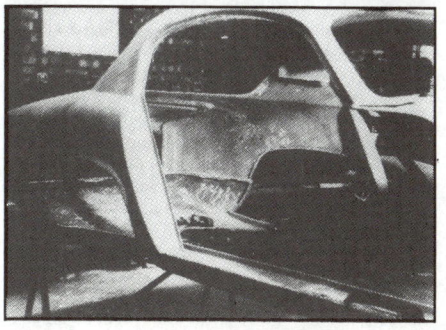

that the alloy casing (a now unobtainable Lotus part containing BMC internals) is bolted tightly against the rubber mountings which provide cushioning against the glass-fibre, but bear in mind that the glass-fibre bodyshell acts as a sounding box, amplifying noise.

On the suspension side, make the usual 'bounce' check for damper condition, and ensure that all the rubber suspension bushes are in good order and the joints tight. Another Elite myth, that the rear Chapman struts can punch their way through the back window, is completely false, although the wishbones and links can be misaligned easily by kerbing. The most likely fault is rumbling from tired rear wheel bearings, generally caused by inadequate greasing and setting of the taper roller bearings. Universal joints last a long time, as long as they have been greased every 500 miles.

The Elite's brakes give no major problems, but it is worth checking that the differential isn't leaking oil over the inboard rear discs, as this can cause seals to perish, fluid to leak and eventually the brakes to fail. The steering should feel light and precise, with no play or vibration. Wear in the steering rack – Alford & Alder on early cars, Triumph Herald on later ones – can be rectified easily.

It is worth considering the state of the interior when examining an Elite as so few trim items can be obtained nowadays. It is quite likely that trim will be worn and split, or unoriginal. Royalite door panels crack easily, but Tony Bates can supply very good replica trim. The Elite's removable plexiglass side windows were prone to leaking, so damp carpets are common. Original carpet is grey, but many cars have a more readily obtainable black replacement. Dampness may have caused the seat runners to rust solid, but replacements can be made by modifying Triumph Herald items.

Small parts were often borrowed from other manufacturers' parts bins – interior door handles are Triumph, exterior handles are from the Commer Cob van or Hillman Husky, and the door locks and much of the switchgear come from the Riley/Wolseley 1.5. A clean and unmarked interior, especially if the original carpets and ribbed grey vinyl spare wheel housing trim are intact, is a good indication of a carefully maintained car.

The roof underside should have a dappled duck-egg blue paint finish, not a headlining. A different colour exterior roof finish – generally silver – is often assumed to indicate a Special Equipment car, but don't take this as gospel. Some cars have gained a two-tone finish during restoration, and a few customers originally specified a single colour for SE models.

None of the Elite's major problems is unsurmountable – just costly. Try to reassure youself that the car you are considering buying is in a reasonable state of health before you ask an expert to confirm your view, or advise on restoration cost. To be honest, Elites are in such demand nowadays that you will have to wait a long time if you're going to be choosy. Their high market value, as well the expense of engine and bodyshell restoration, mean that investing in a professional inspection is vital. Unless you really know what you're looking at, you could easily fall for a speculatively-priced lemon.

PRICES

The sad thing for enthusiasts is that the Elite is now far and away the most expensive classic Lotus, and overseas demand looks set to continue this trend. Some of the best Elites are now going to Japanese, German and Dutch buyers, and we know of owners who have been offered figures exceeding £25,000. But these are the very best cars, so don't lose heart completely.

Around 150 of the 500 or so Elites sold on the British market are known still to exist, but occasionally 'basket cases' are exhumed from barns. Even these, it seems, now command a minimum of £6000, although we have heard of one complete, but dismantled, car selling recently for £9000 because of the number of potential buyers anxious to secure it.

'Middle' money for a running but average Elite should be in the £11,000-£15,000 bracket, but for more than £15,000 you should be looking at a sound car requiring little immediate work. A few years ago no-one would have dreamed that an Elite could cost this much, but its status as a milestone car is now established. Demand now rules the market to such an extent that even experts find it difficult to estimate prices – the answer is to look at as many cars as possible in order to get a measure of the market.

Clubs

Although both **Club Lotus** and **Club Team Lotus** look after the needs of all Lotus owners, there is one club devoted solely to Elite lovers. This is **Club Elite**, whose membership numbers 180, 100 of whom are British. Founded in 1969, it is an invaluable source of advice, knowledge and parts, and offers an immensely useful register (updated annually) of all known Elites. A driving day is held at Goodwood every year, usually on the first Sunday of October, and members are kept informed through an excellent bi-monthly, professionally-printed newsletter. Contact: Nick Raven, Little Questing, 15 Peatling Road, Countesthorpe, Leics LE8 3RD.

Specialists

Fibreglass Services, Charlton Sawmills, Charlton Singleton, Chichester, Sussex (tel: 0243 63320).
A.N.E. Bates, Beech House, East Winterslow, Salisbury, Wilts (tel: 0980 862373).
David Bruzas, Unit 5, Wintonlea, Monument Way West, Woking, Surrey GU21 5EN (tel: 04862 30638).
Climax Engine Services, 82 Northwick Park Estate, Blockley, Glos GL56 9RF (tel: 0386 700631).

Books

Lotus Elite: Racing Car For The Road by Dennis Ortenburger (Newport Press). By far the best historical and practical survey of the Elite, by a highly enthusiastic Lotus man. Lively and superbly illustrated, but sadly long since out of print – it's worth scouring autojumbles for one, although Ortenburger is working on a new Elite book.
Classic and Sportscar Lotus File by Mark Hughes (Temple Press), £8.95. Practical guide, aimed at the novice buyer, covering all first-generation Lotuses from Elite to Europa. Elite has its own chapter. Nicely produced with good colour and black and white photographs, and inexpensive.
Lotus: The Elite, Elan, Europa by Chris Harvey (Haynes), £14.95. A good read and packed full of anecdotes, although these sometimes get in the way of accuracy. About 26 pages on the Elite.
Illustrated Lotus Buyer's Guide by Graham Arnold (Motorbooks International), £9.95. Covers all road-going and some racing Lotuses, so the Elite gets only 10 pages. Fun to read, but not definitive.
Lotus Elite, 1957-64 (Brooklands Books), £5.95. Useful collection of road test reprints in paperback covers.

Acknowledgements

Our gratitude is due to Miles Wilkins and Tony Bates, the two leading Elite specialists, for sharing their knowledge for this feature. David Bruzas also provided much valuable information – he is currently completing restoration of 'the best Elite in the world', which will be the subject of a future C&S feature. Tony Mantle was helpful with engine and transmission information, while David Blacklidge kindly offered his car on a cold winter's day.

OWNER'S VIEW

David Blacklidge fell for the Elite at Earls Court in 1957, but had to wait till 1980 to own one. He explains to Mark Hughes how he brought his sick example back to health

You meet Elite enthusiasts in the strangest places. There we were, parked in a quiet lane near Frome, Somerset, with David Blacklidge's Elite, planning the colour photography you see on the preceding pages, when a woman walking her dog stopped to peer at his car.

"Gosh, an Elite!" she enthused. "We used to have two of those. You belong to Club Elite? I used to put together the club's newsletter 10 years ago. Well, fancy seeing an Elite out here…"

As a doctor, David knows thousands of people in the Frome area, yet here was a local Elite lover he'd never met before. There again, 10 years ago Elite ownership was still a dream for him, a dream he had been nourishing ever since he had admired the first Elite on its public debut at Earl's Court in 1957. He was a Lotus nut even then, and owned a Mk6 which he bought during his student days at Cambridge. He has rarely been without a Lotus since.

The Mk6 served him well, in competition as well as on the road, for two years before being replaced by an S1 Seven. As he lived close to Lotus's Hornsey works, David and his car starred in a Lotus promotional film showing how a Seven is assembled, but he has never seen it! Does anyone have a copy?

David rejoined the Lotus fold in 1968 when he bought an S4 Elan which he assembled himself, and followed that with an Elan Sprint in 1972. "I went to the factory to collect the Sprint, and they put most of the car in the back of my Maxi with a fork-lift truck. I forgot that I didn't have a fork-lift at home, and I needed eight people to help me get the bits out! I became quite good at building up Elans, as I assembled two more for friends. I still have the Sprint – people say the Elan is a heap of trouble, but it is the most reliable car I have ever owned. Its chassis has died now, but I'm beginning to think about the restoration."

To own an Elite, however, had always been David's eventual aim. He hadn't been looking very seriously when he noticed one advertised in a newspaper in 1980. It needed major restoration work, but David felt that he had enough mechanical aptitude to tackle most of the job himself. He made just one journey, back to his garage at home, before taking the Elite to pieces.

"All the mechanical parts – except the accelerator pedal bracket, which wasn't rusty – came off the bodyshell. It took a week just to get the front wishbone bolts out. Apart from minor damage around the nose, the shell was just very old and neglected. It was covered with gel-cracks, one or two of them so severe that I had to cut holes to make sound repairs. I ground out all the gel-cracks and built up the glass-fibre thickness again with mat and resin. There were four layers of paint – yellow, red, and two coats of purple, like an aniseed ball! – which came to ¼in thick in places. Three owners had disregarded the basic glass-fibre rule of rubbing right back before repainting. I had to take the whole lot off, which was exceedingly tedious.

"Some of it came off in great dinner plates, but just removing paint took about 80 hours. It wasn't like stripping a metal car, where you can just slap on the Nitromors and leave it to get on with softening the paint. Nitromors attacks the gel-coat, so you have to put it on small areas and get the paint off within a few minutes.

"Once all the glass-fibre repairs were done, I had to restore the body's smooth profile with filler – I used Plastic Padding's polyester filler. The makers of all these products say they don't sink, but they do. The answer is not to pussyfoot, and put a lot of the stuff on. You leave it for six weeks or so before you sand any off. You really need a wet surface under strip lighting to see the mountains and valleys in the surface clearly, but feeling with your hands is the best way. You literally sculpt the car anew, so to speak. All glass-fibre waves, so you end up with patches of filler all over the place. I tried to obtain the flattest possible finish, but it hasn't lasted perfectly – – glass-fibre is almost living stuff, and a few undulations appeared after a year or so. Someone else did all the painting, using cellulose because you can blend new patches in so well.

"One unusual problem which I've not known other Elite owners to experience, was that the steel hoop on which the doors hinge had rusted badly on one side. This is also the jacking point, and when I tried to raise the car the jack went straight up into the sill. Only the bottom part of the hoop had gone, and the other side was quite all right – the damage was on the nearside, so I can only assume that it was caused by water thrown up from road edges. I had an excellent motor engineer, Dave Phillips, weld in some new steel tube, and a small new section on the front subframe as well. Despite this corrosion the glass-fibre was absolutely perfect, although the car's done 100,000 miles."

Compared with the bodyshell, rebuilding the Coventry Climax engine was quite simple, although parts for it cost £700. It was tired, needing new cylinder liners, piston rings, cam followers and valves, but most of the rest was salvageable. Great attention was paid to balancing, and as a result the car, by Elite standards, is remarkably vibration-free. Again, Dave Phillips did the work which David didn't want to tackle. The car is close to Super 95 Tune, but uses twin SU carburettors instead of Webers.

Parts came together from all sorts of sources. For example, David was unhappy with the tall final drive in his car, but managed to obtain a new 4.2:1 diff when British Leyland discarded some parts stock. He swopped the old diff with someone else who had the Stage II camshaft which he was seeking. He replaced the MG gearbox – "it has awful ratios" – with a rare, but desirable, ZF 'box which Elite specialist Tony Bates happened to have "lying around".

David now covers around 2000 miles a year in his Elite, driving it spiritedly and throughout the year – "the best way to look after a car is to use it regularly". The perfection he sought in his rebuild has paid off with excellent reliability, the one time he has been stranded, on the way to Club Elite day at Goodwood, resulting from a brand new Lucas rotor arm breaking. A passing Lotus Seven came to the Elite's aid.

Although he loves driving his Elite, one suspects that a great part of David's pleasure came from seeing the restoration take shape. After a day spent mending people, he finds mending cars very therapeutic…